Sharing the Love of Gardening
The History of The Westport Garden Club

Sharing the Love of Gardening
The History of The Westport Garden Club

Kristie C. Wolferman

UNIVERSITY OF MISSOURI PRESS

COLUMBIA

Library of Congress Cataloging-in-Publication Data

Names: Wolferman, Kristie C, 1948- author.
Title: Sharing the love of gardening : the history of the Westport Garden
  Club / Kristie C. Wolferman.
Other titles: History of the Westport Garden Club
Description: Columbia : University of Missouri Press, 2025. | Includes
  bibliographical references and index.
Identifiers: LCCN 2024056908 (print) | LCCN 2024056909 (ebook) | ISBN
  9780826223296 (hardcover) | ISBN 9780826275141 (ebook)
Subjects: LCSH: Westport Garden Club (Kansas City, Mo.). | Women
  gardeners--Societies, etc.--Missouri--Kansas City. |
  Gardening--Societies, etc.--Missouri--Kansas City.
Classification: LCC SB403.Z5 W65 2025 (print) | LCC SB403.Z5 (ebook) |
  DDC 635.06/0778411--dc23/eng/20250207
LC record available at https://lccn.loc.gov/2024056908
LC ebook record available at https://lccn.loc.gov/2024056909

Typeface: Adobe Garamond Pro

Contents

Preface

OVER THE YEARS, THE WESTPORT GARDEN CLUB members have made attempts to record the club's history, writing short summaries to turn in to The Garden Club of America. In 2014, Margaret Weatherly Hall and Laura Babcock Sutherland got serious about writing a history. They created an outline and a plan, with the work set to begin in October 2014 and be completed by December 2016. However, club projects always seemed to supersede the time and effort needed to write such a history.

Thus, the concept of this book had taken hold well before I became a member in 2019 and casually mentioned that I thought it would be fun to write the history of The Westport Garden Club. I am so thankful that Margaret Hall agreed to help me. Then came the pandemic, and the Linda Hall Library, where the club's archives reside, closed. Luckily, Margaret had enough material in her basement to get this project started. Since she had been an active member of the WGC since 1988, her knowledge of both the WGC and the GCA proved invaluable. Relying on her own experience, her remarkable memory, and her prodigious collection of photographs and records, she augmented and edited every chapter of this book and wrote all the member biographies, except her own, that are interspersed within the chapters.

I also owe an immense debt of gratitude to club president and professional editor Mary Ann Huddleston Powell, who, during the spare time she did

Figure *Left:* Rose in the Laura Conyers Smith Municipal Rose Garden at Loose Park. (Marianne Maurin Kilroy)

not have, added her special touch to the manuscript. Also, words cannot express my appreciation for the work done by professional photographer Marianne Maurin Kilroy, who spent hours snapping images of members' gardens, culling historical photographs, and editing the hundreds of photos sent to her by club members. The images really do tell the story.

The COVID-19 epidemic of 2020–2022 delayed work on this book, wreaked unprecedented havoc, and affected each of us personally. We had no guideposts from the past to show the way. However, the book went ahead, albeit slowly, and The Westport Garden Club prevailed, with Carolyn Steele Kroh at the helm, followed by Peggy Kline Rooney. Meeting via Zoom allowed our club to carry out its agenda, giving us opportunities to gather virtually regardless of our safe-harbor locations. While the in-person club meetings, fundraisers, flower shows, and workshops temporarily ceased, gardening did not. As in past periods of crisis, our gardens sustained us. Mother Nature, undaunted by human suffering, went about its business of producing a succession of colorful spring and summer blooms, fall colors, and winter respite. We could work in our own gardens, view virtual flower shows and workshops for inspiration, and share images with each other via Instagram or text message. We even had Zoom book clubs and committee meetings. Masked and outside, we still held our wreath workshops and took floral arrangements to hospitals for health care workers.

The GCA also opened new doors to enhance our participation. The national organization made annual meetings with notable speakers available via Zoom to the entire membership—an experience previously open only to attendees.

We are enjoying being back together now with in-person meetings, workshops, and events. Some of these gatherings also include the "men's auxiliary," casually mentioned from time to time in this chronicle. Our men—whether spouses, significant others, or honorary members—play a vital role in the WGC. Besides providing gardening and legal advice, they have been known to cart supplies to workshops, put up with endless time-consuming projects, help build props, create flower arrangements, find great speakers, and even work in their own gardens. The creativity and importance of the "men's auxiliary" lives on.

When WGC members lament that we don't have as much fun as we used to, what are we thinking? Fun is not weighed by random acts of social engagement and uproarious laughter, but by enjoying what we have come together to accomplish. By working together, WGC members have formed deep and lasting friendships and have found peace and inspiration in their gardens, while inspiring others. The members of The Westport Garden Club have also made important contributions and will continue to do so by advocating for civic improvements, producing and conserving food, planting native, and protecting our natural resources and our ecosystem. The educated and committed WGC members are making the world a better place.

As we, the members of The Westport Garden Club, proudly celebrate our seventy-fifth anniversary, we will reflect on all we have accomplished, with this commemorative history book and, of course, with a party!

Kristie Carlson Wolferman

Acknowledgments

With Sincere Thanks to Our Donors

DeSaix "DeeDee" Willson Adams in memory of DeSaix E. Gernes (Willson) and Elizabeth S. Adams

Kimberly Kline Aliber

Elizabeth "Beth" Ritchie Alm

Ellen Bolen

Newell Brookfield

Jill Kathryn Bunting in honor of Margaret Weatherly Hall and in memory of Phoebe Bunting and Diana James

Wendy Hockaday Burcham

Lyndon Gustin Chamberlain in memory of my mother, Margaret "Peggy" Garner Gustin

Lucy Wells Coulson in memory of Norma Sutherland

Cynthia "Cindy" Rapelye Cowherd in memory of Sarah "Sally" Rapelye Cowherd and in honor of Leila Grant Cowherd

Emily Bartlett Darling in memory of Ellen Z. Darling and Eulalie B. Zimmer

Linda S. Evans

Dody Gates Everist

Kathryn Gates in memory of Susie Vawter and Norma Sutherland and in honor of Margaret Hall and Kristie Wolferman

Lorelei Gibson

Elizabeth K. Goodwin

Laura Lee Grace in memory of my mother, founding member Laura Kemper Toll Carkener

Pam Gyllenborg in memory of Mildred Peet Welch and Barbara Welch Thompson

Carlene Hall

Margaret Weatherly Hall in memory of Katherine Buckner Kessinger, Norma Henry Sutherland, and Virginia Haynes Weatherly

Laura Woods Hammond in honor of my mother, Virginia McDonald Miller, and in memory of Jean Holmes McDonald Deacy and Clare Halsell Holmes

Mary Levesque Harrison

Marilyn Bartlett Hebenstreit in memory of Joan Jenkins Bartlett

Susie Heddens

Paget Gates Higgins

Ellen Hockaday in memory of Susie Vawter and Eulalie Zimmer

Blair Peppard Hyde in memory of founding member Ann Peppard White and of Diana James

Alison Bartlett Jager in memory of Joan Jenkins Bartlett

Jennifer Ball Jones in memory of Joy Laws Jones

Nancy Lee Kemper in memory of Mildred Lane Kemper and Laura Kemper Fields

Betty Kessinger in memory of Katherine Buckner Kessinger

Marianne Kilroy

Florence Logan Kline

Carolyn Kroh in memory of Susie Vawter

Lally Family Charitable Fund

Boots M. Leiter

Ginny Bedford McCanse

Virginia McDonald Miller in memory of my mother, Jean Holmes McDonald Deacy, and my grandmother, Clare Halsell Holmes

Jo Missildine in memory of Suzanne Slaughter Vawter

Marsha Moseley in memory of Mildred Peet Welch and Barbara Welch Thompson

Rozzie Motter in memory of my Will and in honor of Jill Bunting

Kay Newell

Marie Bell O'Hara in memory of Jeanne Bleakley

Sharon Wood Orr in memory of my mother, former WGC President Sally Kemper Wood

Ginger Owen

Jack and Susan Pierson Trust

Mary Ann Huddleston Powell

Wendy Jarman Powell

Shelley A. Preston

Ann P. Readey in memory of Alice Parker Scarritt Kelley and Elizabeth Scarritt Adams

Margaret Logan Kline Rooney

Katherine "Kathy" Sawyer

Cacki Smith

Susan Spaulding

Susan Ambler Spencer and Linda Spencer in memory of our mother, Linda B. Spencer, and of our grandmother, founding member Helen Elizabeth McCune Hockaday

Nancy Embry Thiessen

Alison Wiedeman Ward

Sally West

Charlotte White

Illustrations

Image Credits

Front Cover: Herbacious tree peony (*Paeonia lactiflora*). Photo: Lorelei Gibson. This peony came from the Horn/Peet/Gibson garden, documented in the Smithsonian's Archives of American Gardens.

1.2 Wisconsin Historical Society Archives, Elizabeth Abernathy Hull Collection

1.5 Wisconsin Historical Society Archives, Elizabeth Abernathy Hull Collection

1.15 Rozzelle Court before it was enclosed, 1933. Ephemera Collection, RG 70, Nelson-Atkins Museum of Art Archives

3.18 Junior Gallery Plaque, 1960. Westport Garden Club Records, RG 71/01. Nelson-Atkins Museum of Art Archives

3.19 Children trying out the Junior Gallery phones, 1960. Department of Education Records, RG 32, Nelson-Atkins Museum of Art Archives

3.21 Amy Angell Collier Montague Medal won by The Westport Garden Club, 1964. Westport Garden Club Records, RG 71/01, Nelson-Atkins Museum of Art Archives

4.2 *Kansas Prairie*. Photo: Kevin Sink

4.5 *Waving Blue Stem*. Photo: Kevin Sink

4.11 Photo: Kevin Sink

5.11 Photo: Kevin Sink

5.20 and 5.21 "Weatherly Garden," SG-MO08, Smithsonian Institution, Archives of American Gardens, The Garden Club of America Collection

8.2 Harriet Whitney Frishmuth, American (1880–1980). *Joy of the Waters*, ca. 1917; cast after 1945. Copper alloy, 61 x 14½ x 17 inches. The Nelson-Atkins Museum of Art, Kansas City, Missouri. Bequest of Elizabeth A. Hull. F96/38/1 A. Image courtesy Nelson-Atkins Media Services.

9.10 Norma Sutherland's garden, "Heron Haven," SG-KS022, Smithsonian Institution, Archives of American Gardens, The Garden Club of America Collection

9.13 "Weatherly Garden," SG-MO08, Smithsonian Institution, Archives of American Gardens, The Garden Club of America Collection

9.14 "Dillon Garden," SG-KS034, Smithsonian Institution, Archives of American Gardens, The Garden Club of America Collection

9.15 "The Rainbow Garden," SG-KS019, Smithsonian Institution, Archives of American Gardens, The Garden Club of America Collection

9.23 "Heron Haven," SG-KS022, Smithsonian Institution, Archives of American Gardens, The Garden Club of America Collection

10.11 Flint Hills. Photo: Kevin Sink

Introducing the Westport Garden Club

THE YEAR WAS 1950. THAT summer, as Kansas City was emerging from the long war years with renewed vigor, fifty women dreamed of forming a garden club. They were thinking beyond their shared love of gardening. Their goal: to affiliate with the prestigious and influential Garden Club of America (GCA). Therefore, in addition to working to improve their own gardens, they immediately began to learn more about horticulture and conservation. They also engaged in civic projects with the hope that in five years, the minimum amount of time required to qualify, the GCA would invite their club, The Westport Garden Club, to become a member.

When that invitation came in June 1956, the members of The Westport Garden Club (WGC) were excited about the opportunities that association with GCA would afford them. While taking leadership roles in the national organization and contributing to GCA projects since then, WGC members have always put Kansas City and the surrounding region first. They can claim with pride their local work in educational endeavors, conservation, restoration, and beautification. Moreover, through working together on club projects and sharing their love of gardening, WGC members have formed friendships that have stretched across the generations.

Thanks to its dedicated and knowledgeable members, the WGC has played an important role in protecting natural resources and the ecosystem that sustains us all. From its ambitious start in 1950, the club has grown, both in numbers and in influence, to become the vibrant and impactful group it is today.

Sharing the Love of Gardening
The History of The Westport Garden Club

CHAPTER I

The Westport Garden Club Sprouts

IN THE POST–WORLD WAR II era, Kansas City was bustling. Its suburbs were expanding, and the housing boom meant ever more yards with ever more gardens to maintain. In 1950, Kansas City was already home to forty-two garden clubs, so why start another?

The Westport Garden Club began, as do all successful organizations, with an idea: Someone gains inspiration for a project or interest group, then seeks out others who share the same enthusiasm. In the case of The Westport Garden Club, the impetus for its formation came from one Elizabeth Abernathy Hull, of Ridgefield, Connecticut, a colorful character and former Kansas Citian whose family's ties to the area stretched back to before the Civil War.

Described by fellow gardeners as "feisty but also very capable," Elizabeth Hull was a force of nature. She was born in 1900, the daughter of Cora Abernathy and Dr. Albert Gregory Hull, army surgeon and chief adminis-trator of the hospital at Fort Leavenworth. Elizabeth's mother was the daugh-ter of Elizabeth Martin and Colonel James Abernathy. In 1856 Colonel Abernathy founded the Abernathy Furniture Company in Leavenworth, Kansas, the pioneer furniture manufacturer west of the Mississippi. He was also president of the First National Bank of Kansas City. Shortly after his death in 1902, Elizabeth and her family, including her Grandmother Abernathy, moved from Leavenworth to Kansas City, Missouri. Her father began a medical practice there, and the Hulls enrolled Elizabeth, at age five,

Left Figure 1.1 Garden of Margaret Weatherly Hall. (Kilroy)

in The Barstow School, an independent K–12 school. After she graduated in 1917, Elizabeth matriculated at Mount Holyoke College in Massachusetts. In 1919, after her parents divorced, Elizabeth and her mother relocated to New York City.[1]

Elizabeth grew up with an appreciation for nature fostered by both her mother and her grandmother, the latter said to have had "ten green fin-gers." At age twenty, she demonstrated her passion for horticulture when she planned a formal garden at the Hulls' summer home on Madeline Island in Lake Superior. Her three-terrace design included arbors, foun-tains, a screened-in pergola, and 450 rosebushes, among other plantings. In September 1930, *The Kansas City Star* reported that "Miss Elizabeth Hull is becoming famous for her gorgeous garden," which had had a decade to ma-ture by then. In 1936 she and her mother purchased an eighteenth-century house in Ridgefield, Connecticut, which they renovated, and in 1943 they acquired an adjoining twenty-four-acre tract of woods, fields, and wetlands around Silver Spring Swamp. The following year, Miss Hull joined the Ridgefield Garden Club, a Garden Club of America (GCA) member club.[2]

In 1950, Elizabeth Hull's Kansas City friend Helen Hayes Thompson and her husband, Mason, visited her in Connecticut following the gradua-tion of the Thompsons' second son from Princeton. When Helen expressed her enchantment with the cottage garden Elizabeth had designed and im-plemented around her eighteenth-century house, Elizabeth professed that her affiliation with the Ridgefield Garden Club—of which she was projects chair at the time—had taught her a great deal. She had learned not only about gardening and flower arranging but also about conservation, land-scaping, horticulture, and the undertaking of civic improvement projects. Indeed, by this time Miss Hull was also active in the GCA. She served on the GCA Prints and Slides Committee, wrote articles for the GCA *Bulletin*, and acted as a GCA photographer.[3]

Figure 1.2 Boardwalk to the Hull cottage on Madeline Island. (Wisconsin Historical Society Archives [WHSA])

During the Thompsons' visit to Connecticut, Elizabeth lamented that her former hometown of Kansas City did not have a GCA-affiliated club that Helen might join. She encouraged her friend to go home and form one, assuring her that the effort would be well worth her while. It was this idea, of course, that would lead to the foundation of The Westport Garden Club.

Elizabeth Hull certainly knew whereof she spoke. Always very active in her local club, she would not become its president until the age of eighty, when she took it upon herself personally to call upon the members she wanted to serve as committee chairs. When she reached the door of one such member, Shirley Tower, Shirley invited her in. "No," Elizabeth replied. "I am here on business. You will be head of a committee." Shirley agreed, despite having no idea which committee she had just agreed to head. At this, Elizabeth turned to leave, when Shirley asked where her car was. "I walked," was the reply. When Shirley told Elizabeth her husband would

gladly give her a ride home, she responded, "That won't be necessary. I'm capable of walking home," and off she went into the dark night on the mile-and-a-half trip on hilly, winding roads back to her house—at the age of eighty clearly still as "feisty but . . . capable" as ever.[4]

In the town of Ridgefield, Miss Hull, as she was always called, championed efforts "to acquire more open, green space." She was so deeply concerned about the environment and "common-sense horticultural practices" that in 1992, four years before her death, she established and endowed The Garden Club of America's Elizabeth Abernathy Hull Award, in recognition of an individual "who, through working with children under sixteen years of age in horticulture and the environment, has inspired their appreciation of the beauty and fragility of our planet."[5]

Elizabeth Abernathy Hull was an inspiration to Helen Thompson. Along with her words of encouragement to Helen to form a GCA club in Kansas City, she also issued a caveat: In order for her new club to qualify for membership in GCA, Helen would need to assemble a group of fifty women "to work together for five years for civic projects and to improve their own gardens, too."[6]

Implementing the Idea for The Westport Garden Club
Helen Thompson, who had moved to Kansas City from Cleveland, Ohio, as a new bride, might first have entertained the notion of forming a garden group after talking to her hometown friends. Many of them belonged to the Garden Club of Cleveland, a charter member club of GCA founded in 1913. However, it was the trip to Ridgefield that cemented the idea in her mind.

Shortly after her return to Kansas City, in June 1950, Helen invited three friends to join her for lunch, where she planned to propose the idea of forming a garden club—but not just any garden club, since forty-two gardening groups and societies already existed in the Greater Kansas City area—one that could eventually qualify for admittance into The Garden Club of America.

Over lunch on the screened-in porch of Helen's home on Verona Circle overlooking the second green of the golf course of The Kansas City Country

Figure 1.3 *Right:* A profusion of blooms in Margaret Hall's garden. (Kilroy)

Figure 1.4 Inspired by Elizabeth Hull, Helen Hayes Thompson proposed the idea of forming a garden club in Kansas City. (*The Independent*, July 1, 1950)

Figure 1.5 Elizabeth Abernathy Hull (*left*) and her friend Elizabeth Nesbit on Madeline Island in Lake Superior. Liz Nesbit Marty later became a founding member of The Westport Garden Club. (WHSA)

Club, Helen and her guests, Mildred "Millie" Schwartzburg Hoover, Eleanor Nichols "Nicky" Allen (a.k.a. "Ellie"), and Leila Grant Cowherd, had an animated discussion. Reminiscing about the event years later, Millie

Hoover said that she, Nicky, and Leila "were terribly excited about the idea and around that luncheon table on the terrace we worked all afternoon trying to think of fifty women who would really take the idea seriously and not drop out after a few meetings." They may have started with The Kansas City Country Club directory, and that very day the four women came up with a list of possible members. Of course, they included Patti Harding Abernathy, the wife of Elizabeth Abernathy Hull's cousin Taylor, and Elizabeth's childhood friend Elizabeth "Liz" Nesbit Marty. Liz Marty was an established gardener in Kansas City and had won praise for her own Madeline Island garden, designed on property given to her as a wedding gift by Cora Abernathy Hull.[7]

Although not all the potential candidates for the club were tried and true gardeners, all of them were active in civic affairs. The prospective garden club members belonged to various charitable organizations, served on school and church boards, worked for the Red Cross, and were members of the Friends of Art, the support group for the Nelson-Atkins Museum of Art, then called the Nelson Gallery. (The official name was the William

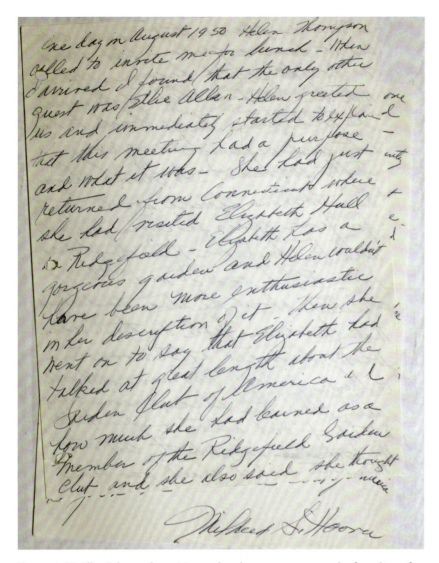

Figure 1.6 Millie Schwartzburg Hoover handwrote a report on the founding of The Westport Garden Club for the club's twenty-fifth anniversary.

Rockhill Nelson Gallery of Art and the Atkins Museum of Fine Arts. The cumbersome name was changed to the Nelson-Atkins Museum of Art in 1982.) Several among the group had artistic talent, such as Marguerite "Margot" Munger Peet. At least one, Ann Peppard White, had a full-time, paying job, something of an anomaly at the time for an upper-class white woman.

Ann Peppard White: Garden Club Career Woman

In 1950, when Ann Peppard White and her sister-in-law Marcella Ryan Peppard joined The Westport Garden Club as founding members, the terms *garden club* and *career woman* were seldom mentioned in the same breath. Annie Pepp, as she was called by her family, changed all that.

Figure 1.7 Despite holding a job outside the home—unusual for married women in the mid-twentieth century—Ann Peppard White found time to spend in her garden. (WGC Archives [WGCA])

Like a number of young women of her era (she was born in 1893.), Ann lost her fiancé in World War I. Looking to support herself, she turned to freelance writing, applying for and landing a job at *The Kansas City Star*, whose editors wanted someone to write about the war from a woman's point of view. Ann's first assignment was to cover a food show, from which she wrote an article about a meat-free "peanut loaf," a creative solution, of course, to the current rationing of fresh meat during wartime. When she took the article to the editor, seated beside him was a large, "distinctive-looking man with very large teeth." As she handed in the piece, she said, "I don't spell very well," to which the large man replied, "Neither do I." He further suggested that he and Ann combine their efforts, saying, "We'll both help win this damn war with your peanuts and my editorials." The man turned out to be former President Theodore Roosevelt, who would write over one hundred wartime editorials for the *Star* in 1917 and 1918. (His last editorial, about the League of Nations, was published after his death on January 6, 1919.)

Ann was hired at a time when there were no female reporters, only a "society editress." She was not allowed in the male-only newsroom and had to submit her weekly column, called "Keep the Home Fires Burning," via a pass-through to a male reporter. Sometimes that reporter was Ernest Hemingway, who had a short career at the *Star*—from October 1917 to April 1918—but who is mentioned in the Ann Peppard White papers that are in the collection of The State Historical Society of Missouri.

In the late 1920s, Ann married James Mayne "Tim" White. Tim's sister, Carmel White Snow, was the editor-in-chief for *Harper's Bazaar* from 1934 to 1958, and his brother, Victor Gerald White, was an artist and muralist. Ann continued to work for the *Star*, writing a children's column and interviewing local notables such as Dr. Katherine Richardson and Ella Loose. She also was a scout for *Good Housekeeping* magazine, suggesting homes and gardens the magazine might feature. Her husband's long illness and death drained her finances, and Ann had to resign from the garden club.

Ann's niece Blair Peppard Hyde, past president of The Westport Garden Club, said that her aunt "never met a person who didn't interest her" and that she was "funny and always positive." Late in life, Ann held salon-like gatherings in her small Rockhill District apartment. Ted Coe, former director of the Nelson-Atkins Museum of Art, said the atmosphere at those meetings resembled that of the gatherings of "the Russian aristocrats in Paris after the Revolution."

The Westport Garden Club has always had a cordial relationship with the press, influenced no doubt by Ann White. ❧

The four ladies divided up the list of names they had generated in order to approach each potential member individually. Millie Hoover related that she and her cohorts were pleased "to find how enthusiastic everyone was over the idea" and decided to convene an introductory meeting.[8]

July 6, 1950, marks the founding of what would be called The Westport Garden Club. In the auditorium of the Pembroke-Country Day School,

Figure 1.8 *Left:* A gate welcomes visitors into Margaret Hall's garden. (Kilroy)

potential members assembled to find out more about the club they had been invited to join. Helen Thompson had asked Millie Hoover to act as president for the first year, so Millie chaired the meeting. The group chose other temporary officers: Helen Thompson would serve as vice president, Nicky Allen as treasurer, and Marcella Ryan Peppard as secretary. Millie and Helen took charge of appointing committees, the first one being the Projects Committee. If the aspiration was to become affiliated with the GCA, the Kansas City group needed to demonstrate its commitment to horticultural and civic projects. Undoubtedly, Helen Thompson took the opportunity at this meeting to explain the history and achievements of the GCA, which at the time had 142 member clubs. She also emphasized that while they were striving to become a member of the national organization, the group's main goal would be to enjoy and share their love of gardening. Nonetheless, it was important to start planning horticultural projects and to find the means to educate themselves on ways to improve their own gardens.

Mildred Schwartzburg Hoover: The Westport Garden Club's First President

By all accounts, Mildred "Millie" Hoover, the first president of The Westport Garden Club, was a force to be reckoned with. A real hands-in-the-dirt gardener, Millie did all her own propagating, weeding, pruning, and cultivating, making her the perfect choice to lead the fledgling club.

Millie's garden occupied an entire lot adjacent to her home and included an apple orchard, a large vegetable garden, and a greenhouse where she raised orchids and started her seedlings. *Mission Hills Magazine* featured her vegetable garden on its Summer 1972 cover along with an article about her organic gardening. (She was way ahead of her time!) Millie was known to ask her dinner guests whether they liked corn, and if the answer was in the affirmative, she would head to her garden to pick corn and then shuck and cook it for that evening's consumption.

Raccoons also found Millie's corn delicious, and an oft-told tale involves her setting a trap one evening to catch the masked bandits. The next morning, still in her nightclothes, she went out to check the

Figure 1.9 Millie Schwartzburg Hoover served as the club's first president. (*The Independent*, April 23, 1955)

trap. It was empty, but she managed to catch herself in the device. Her husband had already left for work—this at a time when cell phones lay in the future—so she had to wait until the postman arrived to be extricated.

Millie was a gifted sculptor who modeled in clay and cast in bronze. She was also a talented pianist who played dual piano concerts with fellow WGC member Virginia Torrance Bolin. A longtime Junior League volunteer, Millie continued to ride her bicycle to work in the Junior League's Thrift Shop until she was in her eighties.

Millie served as GCA zone horticulture representative for three years and received the GCA Club Horticulture Award in 1965. She epitomized the "can-do" attitude that characterized many of WGC's founding members. Tales of her horticultural exploits have acquired near-legendary status with today's members. ❧

Finding a Home at the Linda Hall Library

One of the fledgling group's first decisions was to meet on the first Monday of every other month. (While the club has maintained the tradition of meeting on first Mondays, it now meets every month from September through June.) The introductory gathering of what would be called The Westport Garden Club met at the Pembroke-Country Day School;

subsequent meetings, even up to the present day, would be held at the Linda Hall Library.

The library stands on the fourteen-acre site of the former home of Linda and Herbert Hall. The Halls' will specified that when they died—she in 1938, he in 1941—their large estate and $6 million would be used to found "a free public library for the use of the people of Kansas City and the public generally." Their bequest did not specify the type of library but did stipulate that Herbert Hall's cousin and business partner, Paul Dana Bartlett, should act as the first chairman of the library's board of trustees. With Bartlett at the helm, the Hall trustees researched the kind of library they should establish and decided on a science and technology research library. When the library opened in the Hall home in 1946, it had already gained recognition because of its acquisition of the library of the American Academy of Arts and Sciences, which included important journals by internationally known scientists and first editions of books such as Benjamin Franklin's *Experiments and Observations on Electricity*. The Linda Hall Library would come to be

Figure 1.10 The home of Linda and Herbert Hall originally stood beside the library building, which was completed in 1956. (Linda Hall Library Archives [LHLA])

renowned as "the world's foremost independent research library devoted to science, engineering, and technology."[9]

The Westport Garden Club was able to meet at the Linda Hall Library owing to a most fortuitous connection. One of the WGC's charter members, Julia Jackman Bartlett, was married to Francis Wayland "Frank" Bartlett Jr., Paul Bartlett's brother and a library trustee. Expressing a genuine interest in the gardening group, Frank offered the club the use of a room for their meetings in the Hall library or the Hall home, which remained on the site before being demolished in 1964. In 1975, at the twenty-fifth anniversary of the club, Frank Bartlett's name topped the list of husbands to be thanked, in Frank's case for "permitting The Westport Garden Club to hold their meetings in this beautiful and cultural atmosphere," which was "the envy of every visiting club." Not only has The Westport Garden Club continuously met in the Linda Hall Library, but they store their archives there as well. In addition, members of the Bartlett family have remained on the library board and in the WGC throughout its history.[10]

The Linda Hall Library makes the perfect meeting place for The Westport Garden Club for several reasons. First, meeting in a world-class science and technology library gives The Westport Garden Club credibility as a serious group of gardeners. Second, the library lies in the cultural

Figure 1.12 On the library grounds sits a bench dedicated to WGC member Sallie Bet Ridge Watson. (Kilroy)

Figure 1.11 The entrance to the Linda Hall Library as it appears today. (Kilroy)

center of Kansas City, surrounded by lovely residential areas as well as the University of Missouri–Kansas City, the Nelson-Atkins Museum of Art, the Kauffman Foundation and Memorial Garden, Theis Park, and Southmoreland Park.

Besides the location and its surrounding environment, the grounds of the Hall estate suit the garden club. Herbert and Linda Hall's bequest had not only called for the establishment of a library but also included the wish that the grounds of their former home be preserved and maintained so "that the surrounding trees and grass shall add beauty and dignity to the Library." In the early 1900s, the Halls had expressed their interest in planting unusual and underrepresented trees on their property and had hired the respected local landscape architects Hare & Hare. A variety of trees not native to the Midwest had been nurtured through the years and would continue to be carefully maintained along with new selections. Today this urban arboretum includes more than 300 trees representing 52 genera and 145 species, including 13 trees identified as Greater Kansas City Champion Trees, a designation determined by a mathematical formula based on tree height, spread, and trunk circumference. In 1955, Stanley R. McClane, landscaping superintendent for the J. C. Nichols Company, completed the first survey of champion trees in the Kansas City area. Arborist Chuck Brasher maintained and updated the list from 1974 until his death in 2012, when Powell Gardens took over the responsibility. Apart from the trees, beds of viburnum and tree peonies embellish the grounds, which continue to provide a perfect setting for meetings of The Westport Garden Club.[11]

The first regular meeting of the club at Linda Hall Library occurred on September 11, 1950, in a room in the Hall home. In April 1956 the club began meeting in the new Linda Hall Auditorium on the courtyard level, and since September 1965 the club has met in the present auditorium in what was then a new building on the west side of the library. Forty-nine members attended the September 1950 meeting, each of whom contributed five dollars for annual dues.

Figure 1.13 *Left:* The library is surrounded by trees as the Halls' bequest called for preserving the grounds of their home to "add beauty and dignity to the library." (Kilroy)

What's in a Name?

It was not until the November 1950 meeting that the group decided on a name for their club. They had considered many possibilities, including "The Fledglings," "The Triers," "The Garden Club of Kansas City," and "Heart of America Garden Club," before eventually settling on "The Westport Garden Club." Kansas City itself had begun in the frontier settlement of Westport, so named because it served as an outfitting post on the Santa Fe Trail and the main portal to the West.

1950

Mrs. Wm. Harold Hoover - President
Mrs. Mason L. Thompson - Vice President
Mrs. Earl Allen - Treasurer
Mrs. Joseph G. Peppard - Secretary

1951

Mrs. Joseph Kessinger - President
Mrs. James M. Kemper - Vice President
Mrs. Joseph G. Peppard - Recording Secretary
Mrs. Manvel Davis - Treasurer
Mrs. Byron T. Shutz - Corresponding Secretary

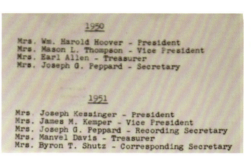

Figure 1.14 Officers of The Westport Garden Club, 1950–1951

In January 1951, The Westport Garden Club decided that the time for temporary officers had expired. A nominating committee presented a slate, which the members approved by unanimous vote. Katherine "Katie" Buckner Kessinger would act as president for 1951–1952; Ruth Rubey Kemper would be vice president until her untimely death in May 1951, when Barbara James McGreevy filled that position; Genevieve Marcell Davis would serve as treasurer; Marcella Peppard would remain on as recording secretary; and Maxine Christopher Shutz would serve as the corresponding secretary. Also in January the first resignation of a member occurred when Mary Dickinson Barton announced to the club that she would soon be moving away from Kansas City. Later she returned as an active member of the club, but her resignation made way for a dynamic new member: Virginia "Ginny" Page Hart. While managing her own antique store, Mrs. Hart would serve as club president in 1957 and 1958 and would help guide the club through many flower festivals.

Affiliation with the Nelson-Atkins Museum of Art

Besides improving their own gardens, the members of The Westport Garden Club had to engage in civic activities if they hoped to receive an invitation from The Garden Club of America to affiliate. Leila Cowherd, the projects chair, and Helen Elizabeth "H.E."McCune Hockaday presented an idea for the club's first civic project, one that would benefit the Nelson-Atkins Museum of Art. They already had talked with the museum director, Paul Gardner, and the Asian art curator, Laurence Sickman, about beautifying Rozzelle Court, then an open-air courtyard inside the museum. Since both Gardner and Sickman were "enthusiastic about a proposed plan," the club voted to install live trees in suitable containers, which could be moved around on wheels within Rozzelle Court. The cost to the club, the Projects Committee had determined, would total $590, and there would be a special assessment of $5 per member, which, obviously, did not cover even half the amount.[12]

Director Gardner had also expressed his desire to have the garden club provide flowers in the museum's period rooms over weekends. The WGC had voted in favor of an additional $5 annual assessment to defray the costs of these arrangements. Then, in April 1951, after Gardner had seen flowers that Newell "Honey Boy" McGee Townley Thornton had arranged for a gathering of Friends of Art members who met monthly for tea on the mezzanine level of the Nelson-Atkins, he asked The Westport Garden Club to create flower arrangements for the May art exhibit of Dorothy Doughty's porcelain birds. Club members, presumably at their own expense, created forty arrangements to complement the exhibit. This first cooperative effort between the museum and The Westport Garden Club opened the door to the club providing floral decorations for the museum's future exhibit openings and galas as well as for Friends of Art events. For example, in fall 1951, the Friends of Art asked members of WGC to provide floral decorations for six afternoon teas and, in 1952 and 1953, for the annual Friends of Art meetings. The relationship between the museum and the WGC "gained increasing significance each year," according to Vivian Pew Foster, but also led daughters and granddaughters of some founding members to think that all WGC members did was arrange flowers for the Nelson-Atkins.[13]

1951 Flood Interrupts First Civic Project

Preparing floral decorations for the museum was the extent of The Westport Garden Club's relationship to the museum for some time, as their first civic project to decorate Rozzelle Court did not happen as planned. Treasurer Nicky Allen called a special meeting to ask the club to cancel the trees for

Figure 1.15 Westport Garden Club members decided that the open-air courtyard at what was then called the Nelson Gallery needed greenery. (Rozzelle Court before it was enclosed, 1933 Ephemera collection, RG 70, Nelson-Atkins Museum of Art Archives [NAMAA])

Rozzelle Court. She had done some research and determined that such trees were unlikely to survive. Instead, she suggested that The Westport Garden Club's first project should be to landscape what was called the Westport Triangle, an area in front of the clubhouse of the Junior League, a nonprofit educational and charitable organization of young women—an organization to which many WGC members belonged or had belonged.

Figure 1.16 The former headquarters of the Junior League sat at the intersection of five major thoroughfares. Landscaping the highly visible front yard proved to be a major undertaking. (WGCA)

Since five important streets come together at the Westport Triangle, this unimproved area of land at Forty-Seventh Street and Roanoke Parkway in the Junior League clubhouse's front yard commanded high visibility. Nicky Allen had already determined the cost of this project at $1,096, which would mean an assessment of $10 per capita instead of the $5 for the Rozzelle Court trees. (These women might have been great at gardening but were not quite as astute at math.) Despite the added expense, the landscaping project passed. Voted on in November, improvement of the Westport Triangle was not slated to begin until spring 1951. In the meantime, however, Mother Nature intervened.[14]

The Great Flood of 1951 was one of the worst natural disasters ever to hit the Midwest. Heavy rainfall began in April and persisted for the next several months, causing flooding in the Kansas, Neosho, Marais des Cygnes, and Verdigris River basins. Rainfall exceeded the annual average by 200 percent in May, by 300 percent in June, and by 400 percent in July. Flooding claimed 28 lives and displaced 85,000 people. More than a million acres of farmland were inundated, and the floodwaters removed the fertile topsoil, leaving behind sediment and sand. One hundred sixteen towns and cities were affected, with 336 businesses destroyed and more than 3,000 flooded. To culminate the rainy season, between July 9th and 12th alone, seventeen to nineteen inches of rain fell, the worst storm in the history of the Midwest. Although the US Army Corps of Engineers spent weeks reinforcing the levees upriver from Kansas City and building dikes, nothing they did prevented the horrific flooding.[15]

Figure 1.17 *The 1951 Flood in Greater Kansas City: A Picture Review*

To many Kansas Citians, July 13, 1951, became known as Black Friday, with much of the city underwater. Located at the confluence of the Kansas and Missouri Rivers, Kansas City owes its place on the map and its early prosperity to its rivers, but those rivers changed the city's history in 1951. Levees and temporary floodwalls gave way, inundating the Argentine and Armourdale Districts in Kansas City, Kansas, swamping the water-processing plant, damaging the railyards, and sweeping people's houses off their foundations. In the flooded stockyards, more than five thousand head of cattle drowned, marking the beginning of the end of the long history of the giant meatpacking businesses in Kansas City, anchored by Armour, Cudahy, Swift, and Wilson. The Fairfax Industrial District, with its petroleum-refinery storage tanks and TWA's airliner overhaul base, lay awash in water and mud, causing TWA to relocate its overhaul operations to higher ground in Platte County. Evacuation orders, sounded at 10 p.m. on July 12, saved many lives, but five Kansas Citians drowned in the deluge. Cleanup efforts could not begin until July 17, after the rivers had crested and the fires that had sprung up from displaced gas and oil tanks had been contained. Even then, many families and businesses, especially those on the West Side and in Kansas City, Kansas, would never return to the area. After the 1951 flood, the Corps of Engineers constructed dams and levees upstream on the Kansas River, leading to the development of many of Kansas's recreational lakes and to more, if not total, control over this kind of flooding.[16]

Figure 1.18 The 1951 flood was one of the worst natural disasters ever to hit the Midwest. Especially devastated were low-lying areas in Kansas City, Kansas.

Like many other residents of Kansas City, those Westport Garden Club members who had running water in the wake of the flood had to boil it for home use and had to endure other personal hardships—flooded basements, marooned yards, and business losses. Nevertheless, just three days after the flood wreaked havoc on the city, WGC members decided to go ahead with social plans they had made for Sunday, July 15. Enid Jackson and R. Crosby Kemper entertained club members and their spouses at their Sleepy Hill Farm, south of Kansas City. As *The Independent: Kansas City's Journal of Society* reported, "A beautiful night in the moonlight took the thoughts of flood-depressed individuals away from their troubles." Three wells provided

Figure 1.19 A garden party at Enid and Crosby Kemper's Sleepy Hill Farm lifted club members' spirits just days after the flooding on Black Friday. (WGCA)

crystal-clear water, and a bountiful dinner and other libations were served on the Kempers' swimming pool terrace.[17]

Despite this needed respite, WGC members certainly could not begin the Westport Triangle renovation, already long delayed by the heavy rains that had started in April. July 1951 marked the first anniversary of The Westport Garden Club, and at the meeting, held just before Black Friday, President Katie Kessinger announced that the Federated Garden Clubs of Missouri had asked the WGC to become a member club. Heated discussion ensued, for some members felt it would be wise to join an established state group rather than wait and hope for an invitation from The Garden Club of America. The majority of the members, however, preferred to wait, and wait they did.[18]

And as they waited, the WGC members began taking more interest in civic issues. In addition to damage done to houses, businesses, and livestock, the city's landscaping and trees had suffered during the flooding. Members of The Westport Garden Club recognized that in the aftermath of the extreme weather, the city was not caring properly for its trees. The city seemed to have employed inept tree surgeons and supposed experts who likely were spraying trees with toxic chemicals harmful to people and the environment. In fall 1951, the WGC fired off letters to L. P. Cookingham, city manager; Martin Chaisson, city forester; and David M. Proctor, city counselor, to protest the damage that the unlicensed tree men were causing. The members asserted that "many trees have been damaged badly in Kansas City by men or crews who pose as experts but are not properly trained." These letters proved effective: The city passed a new law forbidding persons without proper licenses to trim, cut, or prune trees for hire or to treat damaged trees. A *Kansas City Star* article that came out on December 31, 1951, bore the headline "Control on Tree Men: Law Would Require Workers to Have License. At Request of Westport Garden Club, Ordinance Will Be Prepared for Presentation to Council."[19]

Figure 1.20 The WGC wielded influence in its earliest days by urging passage of a new law requiring tree firms doing work in the city to be licensed. (*Kansas City Star*, December 31, 1951)

Back to Business

By the end of 1951, The Westport Garden Club had committees in place and forward-looking plans. At the December meeting, the members finally selected a club emblem, the white dogwood (*Cornus florida*), and passed a motion to make it compulsory for every member to plant at least one white dogwood in her garden. (Almost four years later, on June 20, 1955, the State of Missouri mimicked The Westport Garden Club and made the *Cornus florida* the official state tree.)[20]

In January 1952, the club decided to hold five meetings per year in the Linda Hall Library and five "clinical meetings," actually horticultural workshops, to take place in members' gardens. Professional horticulturists sometimes attended these gatherings, and member participation was mandatory, with every member required to contribute at least once a year to a roundtable discussion of garden topics.

To further their education about gardening, WGC members took field trips. They toured Glendale Farms, near Independence, Missouri, where rosarian Martin Pashea had planted five thousand roses. Pashea shared his tips in a talk entitled "Anyone Can Grow Roses." A spring 1952 trip to the Swope Park Greenhouse inspired club members when they saw the benches of plants and cuttings that would go into city flower beds to regenerate them after the 1951 flood.

The Westport Triangle Project

The Great Flood had caused The Westport Garden Club to delay its plans for the Westport Triangle, so the club would be working on that in 1952 but also decided to undertake a project that year that might help restore some of the flood-damaged areas. Most of the proposed ideas exceeded the expertise and womanpower of the club members, as they considered projects such as landscaping the stockyards or beautifying some part of the Armourdale District. Maxine Shutz, then the WGC projects chair, suggested consulting the Kansas City Park Department superintendent, J. V. Lewis. The superintendent was happy to offer not only his advice but also his assistance. As *The Kansas City Star* reported, "The club, in cooperation with the Park Department, has a program of improving the appearance of various sites throughout the city."[21]

First, WGC members went back to their original plan to landscape the Westport Triangle. Nicky Allen and Margot Peet planned the project, with Eda Marie Peck Luger and Maude Chatten Cowherd helping to flesh out the landscaping details. On April 4, 1952, *The Independent* magazine reported that "Project Number 1 in the ambitious landscaping program of The Westport Garden Club got off to a fine start last week, when members met at the triangle at 47th and Roanoke Parkway in front of the Junior League Clubhouse and proceeded to work." As projects chair, Maxine Shutz had coordinated the project, not only with the Park Department but with the J. C. Nichols Company, the real estate firm that had developed the Country Club Plaza area and residential districts close to the proposed site. Influenced perhaps by the fact that Nicky Allen, J. C. Nichols's daughter, was involved in the planning, men from the Nichols Company helped prepare the ground, delivering black soil to a depth of four feet! A photo in *The Kansas City Star* shows club President Helen Thompson, Georgette Longan O'Brien, Margot Peet, Nicky Allen, Ginny Hart, and Julia Chandler Tinsman "busily at work on their major landscaping effort of the season."[22]

The long-delayed Westport Triangle landscaping finally took shape, although it would not be fully completed until the second planting season of spring 1953, when the garden, conceived of and planted by the WGC, would be turned over to the Junior League, a precedent the garden club would follow with other city beautification projects.

In actuality, landscaping the Westport Triangle proved a much more daunting endeavor than Nicky Allen or most of the other members had envisioned. The initial estimate of $1,096 to landscape the area in front of the Junior League Headquarters had not included labor. Therefore, garden club members did most of the digging and planting themselves, which they should have anticipated. Nonetheless, the project generated a host of complaints about dirty fingernails and aching backs. A frequent refrain was "I'm paying a man to tend my garden while I work untold hours on the Junior League's front yard." A few of the more pampered members, apparently not used to toil, resigned from the project and went so far as to threaten to resign from the club itself, but most of the members enjoyed the hard work

Figure 1.21 *Left:* The white dogwood became the club's emblem in 1951 and four years later, the Missouri state tree. (Kilroy)

Figure 1.22 Attired in sweaters, skirts, and hose—with perhaps the occasional pair of pedal pushers—club members finally began the Junior League landscaping project that had been delayed by the flood. (WGCA)

and camaraderie. In addition, the women learned a lot about gardening from their experience, as well as to be more discerning about the projects they elected to undertake. Years later, Vivian Foster, reflecting on this and the club's other early civic works, reminisced that "the infant garden club was overwhelmingly enthusiastic about doing big things—like landscaping Starlight Theatre or the stockyards after the devastating floods in 1951." It was hard to think small when they considered that GCA, with whom they hoped to affiliate, had engaged in major conservation efforts such as helping to preserve the giant redwood forests of California![23]

Landscaping on Quality Hill

The club's next beautification project, accomplished during the summer and fall of 1952, involved an area downtown at the Missouri Riverfront Park on Quality Hill by The River Club. Lewis Kitchen, one of the developers involved in restoring historic Quality Hill, founding member of

The River Club, and husband of WGC member Charlotte Carnes Kitchen, asked WGC members for advice on planting the areas being rehabilitated after the flood. The club's Horticulture Committee researched what types of plants might grow quickly, add color to the fencing, and be able to thrive in both sun and shade. They selected Paul's Scarlet climbing roses, Japanese honeysuckle, and goji berry bushes.

Paul's Scarlet Climbing Rose (*Rosa* 'Pauls Scarlet Climber'), the most popular of all climbing roses, originated in the United Kingdom in 1915. It derives its name from horticulturist William Paul (1822–1905), noted for his scientific treatise *The Rose Garden*, published in 1903. Paul's Scarlet Climber features bright, crimson-red blossoms, slightly fragrant and very abundant in the spring, and dark green foliage on nearly thornless stems. Unlike other roses, Paul's Scarlet can handle northern exposure and can thrive in both sun and shade, growing to nearly fifteen feet. Per the WGC's suggestions, a professional landscape firm planted fourteen climbing rose plants in November 1952 and planned to install twenty-eight additional Paul's Scarlets in the spring.

Figure 1.23 The club chose Paul's Scarlet climbing roses to add color to the rehabilitated landscaping on Quality Hill near The River Club after the flood.

Purple-leaf Japanese honeysuckle (*Lonicera japonica* 'Purpurea'), a vigorous evergreen vine that can reach from ten to over thirty feet tall, also is a hardy plant that can tolerate part shade. Its fragrant flowers are followed in late summer by blue-black, berry-like fruits that attract birds. Although this species of honeysuckle is now considered invasive and therefore undesirable,

the Horticulture Committee selected this plant because of its rapid growth, its fragrance, and its beauty; they supervised the planting of twenty-five purple-leaf Japanese honeysuckles.

Figure 1.24 Plantings at The River Club included purple-leaf Japanese honeysuckle and goji berry bushes, chosen for their hardiness and rapid growth. (WGCA)

Dominating the Quality Hill planting were the 110 goji berry, or Chinese wolfberry, bushes (*Lycium chinense*) placed along the fence south of The River Club between Jefferson and Pennsylvania. A deciduous wood shrub tolerant of a wide range of soil and climate conditions, these bushes grow to a height of three to nine feet and produce goji berries, or wolfberries, which also attract birds. In a thank-you note written to WGC President Katie

Kessinger, Lewis Kitchen stated, "I am sure the result will be very beautiful, and I do want to express my appreciation to you and the other 'gals' on the committee for your help and assistance."[24]

At the twenty-fifth anniversary celebration of the WGC in 1975, Vivian Foster reported that the Paul's Scarlets still bloomed on the fences on Quality Hill, and that the triangle in front of the Junior League Headquarters still shone as an early WGC achievement. Along with these two projects, also in the summer of 1952, The Westport Garden Club donated ten white dogwoods, their emblem tree, to the grounds of St. Luke's Hospital. Unfortunately, those trees no longer stand, having been swallowed up by excavators' shovels during a hospital renovation.[25]

Incorporation

On September 10, 1953, The Westport Garden Club passed another important milestone toward becoming a member of The Garden Club of America when it received its Certificate of Incorporation from the Office of the Secretary of State of Missouri, becoming the third organization incorporated in Missouri under the state's new Not-for-Profit Corporation Law. Thanks to George L. Gordon, husband of WGC member Jane Hemingway Gordon, the law office of Gordon and Gilmore handled the paperwork required for the application. In a letter to club President Maxine Maxwell Goodwin, George Gordon laid out the steps necessary to go from an unincorporated to an incorporated club. First the board of directors (the language used in the law) and then the entire membership needed to meet to approve the articles of incorporation and the bylaws to be submitted with the application. The club needed to inform its bank of the new status and, thereafter, would have to submit an annual report to the Office of the Secretary of State. Gordon also offered the women a word of caution: Should the club ever wish to amend the original articles of incorporation, the WGC would have to follow a prescribed statutory procedure and file a certificate of amendment with the secretary of state.[26]

In 1953, the possibility of ever changing the articles of incorporation seemed remote, but ten years later, in March 1963, club President Vivian Foster filed a certificate of amendment. The desired change regarded a paragraph referring to the club's purpose in article II and the addition to article IX of a plan of distribution of assets in the case of dissolution of the

Figure 1.25 Certificate of Incorporation

club. Based on his reading of the changes in article II, the Internal Revenue Service district director was not sure the club should retain its tax-exempt status. He thought the new language implied that the club might not dedicate itself exclusively to charitable and educational purposes, as required for 501(c)(3) status. Therefore, he referred the amended application to the IRS in Washington, DC. George Gordon's office, which again helped with the application, told Vivian that if the action was unfavorable, he and his colleagues "would endeavor to do all we can to secure final approval." In fact, the Washington, DC, IRS office issued an unfavorable ruling on July 28, 1964, stating that The Westport Garden Club was no longer entitled to tax-exempt status owing to the change made to one of its purposes delineated in article II. That, however, was not the end of the matter. The IRS did not realize what it was up against when it tested the women of the WGC and the men in their "men's auxiliary," many of whom sent an onslaught of letters to Washington, DC, along with evidence of their charitable and educational purposes. Apparently, something in the letters, if not the great number of them, had an effect. In October, Lawrence B. Jerome, chief of the Washington, DC, Exempt Organization Branch of the IRS, personally revoked the previous ruling himself and returned The Westport Garden Club to tax-exempt status. His letter stated, "Based on additional information submitted with your letter of protest dated September 16, 1964, and the additional information, including copies of the Amendment to your Articles of Incorporation, furnished with your letter of September 22, 1965, we have now concluded that you are both organized and operated exclusively for charitable and educational purposes within the contemplation of section 501(c)(3) of the 1954 code."[27]

Meanwhile, The Westport Garden Club, having regained its tax-exempt status, continued to undertake other civic projects, if none quite as strenuous as the Westport Triangle. In 1954, the club planted flowering crabs at St. Luke's Hospital. They did foundation planting and added shade trees around Grant Hall at the Kansas City Conservatory of Music. In the Linda Hall Library's west courtyard, WGC members planted a "green garden," designed by Maxine Goodwin, and installed the lead facing for a pool and a lead figure of Mercury sculpted by Wheeler Williams. These were just the first of many gifts WGC presented to Linda Hall Library in appreciation for the privilege of meeting there.[28]

Flower Shows Blossom

Perhaps The Westport Garden Club's greatest inspiration in its early years was the concept of one day hosting a flower show at the Nelson-Atkins Museum of Art. In November 1954, several members of the Projects Committee took this idea to Laurence Sickman, who had been appointed director of the museum in May 1953 after Paul Gardner retired. Sickman expressed

Figure 1.26 St. Luke's Hospital was the beneficiary of two tree plantings in the club's early years: ten white dogwoods in 1952 and flowering crabs two years later. (WGCA)

his support for the proposed show but, of course, needed to get approval from the museum's de facto governing board, the University Trustees—men appointed by the presidents of the Universities of Kansas, Missouri, and Oklahoma, this according to William Rockhill Nelson's will. These trustees at the time were Robert B. Caldwell, Milton McGreevy, and David T. Beals Jr. It had to have helped the WGC cause that Milton McGreevy's wife, Barbara, and David Beals's wife, Helen, were WGC members. In fact, Helen Ward Beals would act as co-chair of the proposed flower show, with Eda Marie Luger. Even given the connections, the trustees were reticent to add another social event to the museum's busy schedule, as art and education remained their top priorities. However, they had already agreed to host the Jewel Ball, a debutante ball, at the museum, the first such event having occurred with much fanfare in June 1954. Therefore, at their meeting on December 2, 1954, the museum trustees approved The Westport Garden Club's proposal to hold a flower show in May 1955. They decided that since "this organization is composed of some 50 prominent young matrons and most of them are members of the Friends of Art," the garden club "would bring an increased number of desirable people into the Gallery."[29]

The First Flower Festival: 1955

WGC members wanted their first flower festival, loosely patterned on a Montreal show called Fête des Fleurs, to be a smashing success; it would be the first flower show in the United States staged in an art museum. Marjorie Lane Paxton used her artistic talents to help design posters and brochures for the festival. Other club members sent announcements of the May 20, 1955, show to all ninety-three of the garden-related groups in the Greater Kansas City area and went to talk to the members of the larger clubs, encouraging them to participate in the festival. Although eighteen clubs responded favorably, only four other clubs actually mounted displays at the festival: the African Violet Club of Greater Kansas City, the Missouri Branch of the American Begonia Society, the Gloxinia Society of Greater Kansas City, and the Orchid Society of Greater Kansas City.[30]

It was decided that both members and nonmembers of The Westport Garden Club could submit flower arrangements for the festival, all of which would be judged in a competition that included eight different artistic classes, including one specifically for professional florists. At the event, judges from Washington state; Denver, Colorado; and St. Joseph, Missouri, determined the best flower arrangements in each artistic class as well as five horticultural classes, which consisted of flowering shrubs, hybrid tea roses, floribunda roses, peonies, and perennial flowers. Besides the clubs' display of floral arrangements and horticultural specimens, the museum arranged a special flower-related exhibition of artwork from Europe, the Americas, and the Far East. To make the WGC Flower Festival even more special, artist Thomas Hart Benton donated a miniature still life painting to be awarded to the winner of a drawing. When it was all said and done, Katie Kessinger proclaimed that no other city could possibly have "such a glamorous, perfect setting for a Flower Show."[31]

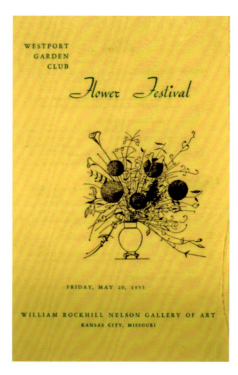

Figure 1.27 The club's first flower festival, held at the Nelson-Atkins Museum of Art, was also the first flower show in the United States to be displayed in an art museum. (WGCA)

Figure 1.28 Marjorie Lane Paxton helped design the posters advertising the festival. (*The Independent*, May 28, 1955, WGCA)

The 1955 Flower Festival did indeed prove to be a great success, but it was quite an undertaking, one pulled off only with the help of many contributors and museum personnel. In the planning stages, The Westport Garden Club had agreed to split the money raised from the raffle tickets and the price of admission with the Nelson-Atkins. When the net proceeds amounted to only $364.08, Katie Kessinger, on behalf of all the garden club members, apologized to the museum, considering all the work and help given by its staff. Sickman replied, "It isn't the money that is important to us, though of course we need and are deeply grateful for it, but we appreciate even more the fact that some of the people who attend your shows would otherwise never step in our doors." Besides giving half the proceeds from the festival to the museum, the club voted in June 1955 to return to the idea of decorating Rozzelle Court, this time allocating $1,400 "to beautify" the space.[32]

The first festival not only had cemented a connection between The Westport Garden Club and the Nelson-Atkins Museum of Art but also had promoted a more meaningful connection between the WCG and the whole Kansas City horticultural community: All gardeners and garden clubs in the Greater Kansas City area had been invited to participate in the civic event. For its part, the museum agreed to host the club's future flower festivals. The favorable publicity portended well for both the WGC and the Nelson-Atkins Museum of Art.

The Second Flower Festival: 1956

The Westport Garden Club held its second flower festival in May 1956. Katherine "Katy" Fisher Satterlee and Martha Callaway Knight served as publicity co-chairs. They arranged for many pre-festival radio interviews and television appearances to bring the 1956 Flower Festival to the public's attention. On the television show of one spot-on local celebrity, Bette Hayes, which aired every afternoon on WDAF, event Co-Chairs Katie Kessinger and Marcella Peppard discussed the collaboration between The Westport Garden Club and the Nelson-Atkins with Museum Director Sickman.

As with the first festival, the WGC invited other garden and garden-related clubs to contribute. Although they once again contacted ninety-three such clubs, only nine of them entered. The Orchid Society staged a beautiful exhibit, but some of the other participating clubs brought their

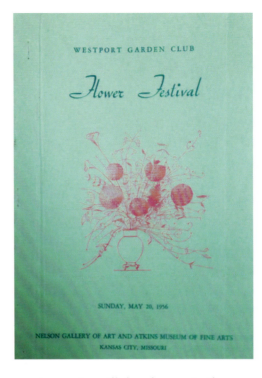

Figure 1.29 The second flower festival, held in 1956, was described in the May 26 *Independent* magazine as "a brilliant production."

specimens in milk bottles or similar containers, much to the consternation of WGC members. They realized that in the future they should run the show themselves without expecting help from other clubs.[33]

Flower arrangements reflecting the 1956 theme "Gracious Living" filled Rozzelle Court. Club members and friends entered compositions in a variety of classes, including monochromatic, miniature, horticulture, and design. For this last category, twenty pedestals, placed on large columns in Kirkwood Hall, displayed the entries. The winning entry was made by Elizabeth "Libby" Scarritt Adams, who was not yet a WGC member but soon would become one. A highlight of this second festival was a demonstration whereby each of nineteen entrants made an arrangement with a select group of flowers and a container donated by the museum. *The Independent* (which did not use women's first names until well into the 1970s) reported that the event co-chairs, Mrs. Kessinger and Mrs. Peppard, "lent their talent and energy for the breath-takingly beautiful display" and that the show itself was "a brilliant production that made Kirkwood Hall, Rozzelle Court, and the Loan Rooms a bower of blossoms and a riot of color."[34]

THE KANSAS CITY STAR.

KANSAS CITY, APRIL 21, 1957—SUNDAY

Flower Festival Recalls Rozzelle Court's Donor.

By Winifred Shields

AN inviting arrangement of a garland of magnolia leaves on the ancient fountain in Rozzelle court at the Nelson Gallery of Art for the Westport Garden club's annual flower festival Thursday, will bring renewed interest to this architectural feature.

It has been observed before that in Rozzelle court the seasons never change. The fountain goes off in the winter and is turned back on in the spring. But except for this acknowledgment of passing time, and what you can learn from the warmth of the sun or the intensity of the sky, there is nothing to tell you what season it is.

From Ancient Roman Villa.

If you linger very long you may not even be sure of the century. The fountain basin is a Roman bath from Hadrian's villa at Tivoli. It is almost 2,000 years old. The marble sarcophagus under the cloisters on the north wall is Early Christian. The School of Della Robbia plaques and the carved stone emblems of once-powerful Italian families affixed to the walls above and below the balconies are 15th century. So is the architectural style of the court, itself.

The courtyard was the inspiration of the late Thomas Wight, architect for the Nelson Gallery of Art, who designed many Kansas City buildings. Wight was devoted to the principles of classical architecture but hated anything that looked brash, new or hard. When he commissioned Daniel MacMorris, artist to paint ceiling decorations in this Renaissance court at the gallery he put it into the contract that MacMor-

-ris must make the paintings look "500 years old." He then supplied MacMorris with his own copy of a book of steel engravings of Raphael decorations in the Vatican.

A Task of Many Months.

With this for reference MacMorris went to work. Eighteen months later he had finished the arduous labor of decorating the vaulted ceilings of the court in Renaissance style. The painting was done, for the most part, in 1938. The

court had been constructed at least five years earlier, but Wight was in no hurry to have the decorations put on. He wanted the materials to "mellow" first. In the meantime the architect had MacMorris studying the colors of old Persian paintings in the archives of the Metropolitan Museum of Art.

The court is not a copy of any particular Italian court, but a composite of several that Wight had seen and liked.

While Thomas Wight gave

Rozzelle court its form, it was the late Frank F. Rozzelle, lawyer, who gave it being. Attorney for William Rockhill Nelson, founder of The Star, and the Nelson family, Rozzelle provided the money with which the court was built. After drawing up the numerous Nelson wills, leaving a newspaper fortune for the establishment of an art museum, Rozzelle wrote his own making a similar disposition of his fortune. He left $200,-

000 almost his entire estate, to be used for building.

If perpetuating his memory is what Rozzelle had in mind his name has, in a sense, been a disservice to him. "Rozzelle" court has a decorative sound but does not at first suggest an individual. Thousands of persons enjoy the beauties of Rozzelle court every year but it is unlikely that many of them know anything about Mr. Rozzelle.

Mr. Rozzelle's Activities Here.

Born in Scott County, Kentucky, in 1857, Frank F. Rozzelle grew up in Missouri on a farm in Caldwell County. He entered the University of Missouri when he was 15, was graduated from there, and went on to law school at the University of Michigan. On his way home from law school he bought a railroad ticket to Kansas City, liked it here, and stayed.

A member of the law firm of Rozzelle, Vineyard & Boys, he was outstandingly successful in his profession. He considered himself too shy to make effective pleas in court, however, and was never an outstanding trial lawyer.

Mr. Rozzelle was police commissioner in Kansas City under Governor Folk from 1905 to 1907. One of his acts in this capacity was to close saloons on Sundays. He was also a president of the University club.

A bachelor, Mr. Rozzelle built for himself a handsome home on Main street just north of Armour boulevard and entertained handsomely.

He is remembered for sometimes hiding his own personality behind those of his associates. It has been suggested that Rozzelle court is hidden in the gallery in much the same way.

A WREATH OF MAGNOLIA LEAVES WILL GARLAND THE FOUNTAIN IN ROZZELLE COURT of the Nelson Gallery of Art for the Westport Garden club's annual Flower festival there this Thursday. The club recently had installed in the court stone copings to contain plant beds in each of the corners. The club also has provided the plants and shrubs. The festival will combine the showing of flower arrangements with displays of old silver and porcelain. Shown planning the fountain decorations are Mrs. Milton McGreevy, left, 5707 Oakwood road, Mission Hills, and Mrs. Frank Paxton, 5600 Pembroke lane, Mission Hills.

Figure 1.30 *The Kansas City Star* found each of the WGC flower shows newsworthy. This article highlighted the third festival, in 1957. (*Kansas City Star*, April 21, 1957)

As with the first festival, the club sold raffle tickets for a dollar each, and the prize that year was a silver-point drawing by artist Frederic James. Although the net profit to the club this time around was $1,615, the club did not share the profit from the raffle tickets, so the Nelson-Atkins received only half of the remaining net profits, or $239.55. True to form, Director Laurence Sickman remained grateful for whatever money the WGC could add to the museum's unrestricted funds, saying that the flower show won "for the Gallery a great deal of good will." In his annual report, Sickman

remarked that two activities, the Jewel Ball and the WGC Flower Festival, "add special lustre to the schedule of yearly events" and provide "favorable publicity," which highlights "the Gallery's contribution to the civic and cultural life of Kansas City." Later that same year, The Westport Garden Club sponsored a special lecture at the museum: "English Gardens and Country Homes," thereby bringing more people to the museum who might not otherwise visit.[35]

GCA Membership at Last

The Westport Garden Club, which had begun in 1950 and had incorporated in 1953, decided at its annual meeting in June 1955 that it was time—after five years in the growing and learning phase—to submit the club's application to become a member of The Garden Club of America. Although there were still members who preferred joining the Federated Garden Clubs of Missouri (part of the National Council of State Garden Clubs, Inc.) and others who just wanted to garden and were not concerned with joining a national organization, the majority voted to become affiliated with GCA should their efforts qualify. Mrs. Alfred Kiekhafter, Millie Hoover's aunt and a member of Green Tree Garden Club of Milwaukee, proposed the WGC for membership. Mrs. J. Churchill Owen of the Garden Club of Denver seconded the proposal. Mrs. Owen, a longtime friend of Helen Beals's, was an accomplished GCA floral design judge and had come to Kansas City as Mrs. Beals's guest to judge the entries in the first flower festival. (Mrs. Owen would go on to become a director on the GCA Board, beginning in 1956.) An additional second to the WGC proposal came from none other than Ridgefield Garden Club member Elizabeth Abernathy Hull. In June 1956, Mrs. Charles D. Webster, GCA corresponding secretary, informed WGC President H.E. Hockaday that The Westport Garden Club had been accepted into membership of the national organization. ❧

CHAPTER 2

The Garden Club of America: Strength in Numbers

WHEN THE WESTPORT GARDEN CLUB became affiliated with The Garden Club of America in 1956, the national organization already had a long and significant record of accomplishments in the fields of horticulture, conservation, and beautification. The ladies who founded the GCA in 1913 had first been drawn together by their love of gardening, but they included in the national club's objectives both civic planning and the protection of native plants and birds.

At that time, urban planning and landscape architecture were in their infancy; both conservation and preservation were basically unheard of, and women, who did not even have the right to vote, could only hope to bring about environmentally protective political change through their husbands, brothers, or fathers. Nevertheless, GCA members proved to have remarkable stamina when it came to promoting their ideals for preserving and enhancing the American landscape. Like other progressive-minded clubwomen of the era, even without enfranchisement, the members of GCA made their voices heard and helped effect the passage of meaningful legislation on the state and national levels. They not only set the standard for beautiful gardens but shaped the world they lived in through their conservation efforts: restoring the grounds of historic estates, advocating for the preservation of wildflowers and trees, re-creating the era of old-fashioned flower gardens, and preserving and expanding green spaces in a rapidly changing urban environment.

Figure 2.1 *Left:* A bee shows interest in a lotus in Blair Peppard Hyde's water garden. (Kilroy)

Establishing The Garden Club of America

Although the Ladies Garden Club of Athens, Georgia, dating from 1891, claims to be the oldest women's garden club in the United States, the Garden Club of Philadelphia, conceived in 1904, is among the nation's earliest garden clubs. It provided the model and inspiration for The Garden Club of America. With the motto *Furor Hortensis*, translated as "Garden Mad," the objectives of the Garden Club of Philadelphia were "beautification, conservation, preservation." Through fundraising projects, such as flower shows, club members were able to provide financial support for arboretums and historic gardens in the area and promote interest in garden design and maintenance, conservation projects, and horticultural research.[1]

Two remarkable women, Elizabeth Price Martin and Ernestine Abercrombie Goodman, conceived the idea of the Garden Club of Philadelphia and, only nine years later, The Garden Club of America. Mrs. Martin, the wife of a distinguished judge, and Miss Goodman, eight years younger than her cohort, were both from Philadelphia and had summer residences in Chestnut Hill, where they shared a mutual love of gardening and a respect for history. Given the success of the Garden Club of Philadelphia, the two women, along with their friend Elizabeth Bayard Henry, had the idea of calling together members of other garden clubs that shared a similar purpose and enthusiasm for gardening. They wanted to establish a national organization "in order to achieve larger civic goals more efficiently." They also assumed that local clubs would continue to address their respective "community needs in horticulture, conservation, environment, and related

fields" in accordance with the Garden Club of Philadelphia's mission statement. Thus, it is the Garden Club of Philadelphia that is recognized as "having sown the seed which has sprouted far and wide into more and more garden clubs, all working for the beauty of America."[2]

Figure 2.3 Elizabeth Price Martin, a founder of both the Garden Club of Philadelphia and The Garden Club of America, served as president of GCA from 1913 until 1920. When she died in 1932, flags flew at half-mast in Philadelphia. (Garden Club of Philadelphia Archives [GCPA])

In the spring of 1913, the women of the Garden Club of Philadelphia invited representatives from eleven other known and similar garden clubs in Illinois, Maryland, Michigan, New Jersey, New York, Pennsylvania, and Virginia to gather in Philadelphia for the privilege of founding a national garden club. According to Ernestine Goodman, "There was enthusiasm, there was indifference, there was consternation among those who feared that being organized would destroy our independence and kill the joy of living." Such concerns aside, the meeting proved to be a success, beginning on April 30, 1913, with a luncheon for twenty-four ladies hosted by Elizabeth Henry in Germantown, Pennsylvania. The following day, at nearby Stenton,

Figure 2.2 *Left:* GCA set the standards for formal gardens such as Kauffman Memorial Garden in Kansas City. (Kristie Carlson Wolferman)

Figure 2.4 Co-founder of both the Garden Club of Philadelphia and the GCA, Ernestine Abercrombie Goodman wrote the GCA's mission statement that served the organization well into the 1970s. (GCPA)

the original home of James Logan, secretary to William Penn, the assembled representatives of the twelve clubs agreed to join together in a national organization and invited four other garden clubs to become members of The Garden Club of America.[3]

The twenty-four representatives chose officers, who would form an Executive Committee. Not surprisingly, Mrs. Martin was elected president of GCA and Miss Goodman, secretary-treasurer, while the office in Mrs. Martin's Rittenhouse Square home in Philadelphia would become the

Figure 2.5 Representatives from twelve garden clubs gathered in Germantown, Pennsylvania, at Stenton, the original home of James Logan, secretary to William Penn.

headquarters of the new organization. Four vice presidents from other clubs rounded out the Executive Committee: Helena Rutherford Ely, from the Garden Club of Orange and Duchess Counties in New York; Albertina Pyne Russell, from The Garden Club of Princeton, New Jersey; Louisa Yeomans King, from the Garden Club of Michigan in Grosse Pointe; and Katherine "Kate" Lancaster Brewster, from the Lake Forest Garden Club in Illinois.

The conference attendees also approved a mission statement that, according to what may be an apocryphal story, Miss Goodman had been sent out of the room to write. When she returned, she had put together the words that would define the GCA: "The objects of this association shall be to stimulate the knowledge and love of gardening among amateurs; to share the advantages of association through conference and correspondence in this country and abroad; to aid in the protection of native plants and birds; and to encourage civic planning." Having established their objectives in a mission statement that would not be updated until the 1970s, the ladies listened to Letitia Ellicott Carpenter Wright, who had re-created the Colonial Garden at Stenton, deliver a paper on "horticultural matters of the day." They then toured Penshurst, the amazing gardens of Bessye Roberts, setting the tone for future annual meetings: lovely luncheons, business meetings in historic settings, informational lectures, reflections on how to meet the club's objectives, and tours of beautiful gardens. The GCA had been launched.[4]

A GCA Publication: The *Bulletin*

Late in May 1913 Elizabeth Martin assembled a committee of four to write the GCA constitution. Letitia Glenn Biddle, one of the committee members, invited the others to meet at one of America's great historic houses, Andalusia, her husband's family home on the Delaware River. After drafting the constitution, they also agreed on the creation of a GCA publication. Conceived and compiled by Mrs. Martin, as "the means of bringing into closer touch the Clubs composing 'The Garden Club of America,'" the first *Bulletin* was a four-page magazine issued that spring. By the time of the third annual meeting of the GCA in 1915, Kate Brewster had become the sole editor of the now twelve-page *Bulletin*, taking on the job of editing, paying for, and expanding the magazine for the next six years, suspending

publication only from May 1918 to November 1919, when she served in a war capacity. Being editor of the *Bulletin* was a full-time job. "It may be well to remind you," Kate Brewster wrote in the July 1920 *Bulletin*, "that the editor enjoys the *Bulletin* much as a mother does a troublesome pair of twins. . . . Impatience, affection, exhaustion, and devotion are all implied." The *Bulletin* became the main, but not the sole, publication of the GCA, and is today published quarterly.[5]

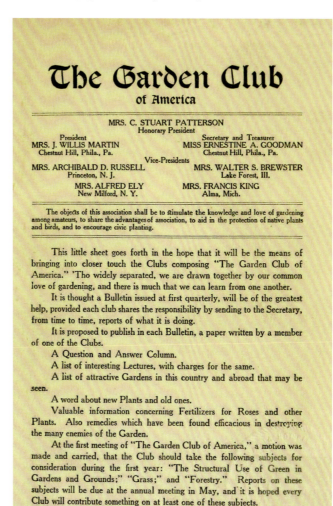

Figure 2.6 Elizabeth Martin conceived the idea of a publication, and Kate Brewster became the editor of the GCA *Bulletin* in 1915. Editing the new magazine became a full-time job. (GCPA)

The April 1914 issue of the *Bulletin* included information on the first meeting of the Council of Presidents, held at the New Jersey home of Albertina Russell preceding GCA's second annual meeting in May in Princeton, New Jersey. Five new garden clubs had been admitted to GCA, raising the total to twenty-one, each of which sent two voting delegates to Princeton. This meeting was attended by over one hundred GCA members, "including non-delegates, who were most heartily welcomed." The delegates put into motion a bevy of new garden club ideas: a year-long study on the placement of flower gardens in landscaping, a committee dedicated to beautification of highways, and the goal of creating a color chart—a set of standards for colors and color combinations for use in both flower arranging and distributing plants in gardens.[6]

The GCA Springs into Action During World War I

Some of these early endeavors had to be postponed, however, owing to the outbreak of World War I. Almost as soon as the first shots were fired, killing Archduke Franz Ferdinand and his wife, Sophie, on June 28, 1914, the infant GCA sprang into action. Although flower gardens and beautification remained on their agendas, club members began writing booklets on growing vegetables. Also, the secretary-treasurer of the GCA, Ernestine Goodman, strongly requested the United States Department of Agriculture "to issue a Special Bulletin . . . so that measures may be taken . . . to preserve domestic seeds, as there may be a shortage of foreign seeds next year."[7]

The GCA continued to meet during the war. At the third annual meeting, in May 1915 in Baltimore, eight new clubs were admitted to the national organization, followed by three more clubs in 1916. During the summer of 1916, the GCA formally began its work in conservation with the formation of the Wildflower Preservation Committee.

In 1917, when the United States formally entered the conflict already raging in Europe, GCA President Elizabeth Martin called a special meeting. "America is at war," she announced, "and The Garden Club of America must assume what responsibility its name implies," suggesting that the gardens of America focus on food production. "We stand or fall by our food supply," Kate Brewster added. When Secretary of War Newton Baker established a Women's Committee of the Council of National Defense to coordinate women's war work, Elizabeth Martin volunteered herself for the committee

and went to Washington, DC, to meet with other leaders of Progressive-era women's clubs, including Louisa King, president of the Women's Farm and Garden Association (WF&GA). Both Mrs. Martin and Mrs. King, who was also a founding member and vice president of GCA, agreed that GCA and WF&GA should lead the way. They would create a Woman's Land Army of America (WLAA), modeled after England's Woman's Land Army, which trained and encouraged young women to take the place of men in the farm fields of home. For that purpose, in December 1917, these two women, along with GCA members Delia West Marble, Ernestine Goodman, and Anna Gilman Hill, headed to Washington, DC, to meet with Labor Secretary William Bauchop Wilson. Despite the obvious cultural difference between these well-heeled (and unenfranchised) women and Secretary Wilson, who had grown up with his coal-mining parents and nine siblings near the Pennsylvania coal mines, they reached an understanding. On the spot, Secretary Wilson made the WLAA a new agency of the agriculture division of his Labor Department. Immediately, the GCA set to work establishing agricultural courses at various colleges and universities. The Garden Club of Philadelphia even funded expenses for European girls to come to the United States and work alongside American college students to learn the basics of vegetable gardening, skills that would prove vital in war-torn Europe. Meanwhile, Delia Marble, encouraged and sponsored by her father, Manton Marble, who owned the *New York World* newspaper, took to the rails, traveling the country to organize and promote the WLAA, eventually recruiting fifteen to twenty thousand "farmerettes" in forty states and the District of Columbia. In the planting and maintaining of so-called Victory Farms across the United States, members of GCA provided both monetary and physical support for food production.[8]

In the July 1917 issue of the *Bulletin*, Mrs. Martin emphasized "the serious part that Garden Clubs of this country will have in the production and conservation of food for the Nation." However, she also urged GCA members not to give up their flower gardens. "Let us find in our gardens rest, peace and inspiration. . . . Keep up the beauty of your gardens if only in simple annuals—not only for yourselves but for all who may be able to share them with you. Work in your gardens on Saturday, rest in them on Sunday, and you will be better fitted on Monday for the services you are giving to your country." In fact, the gentlewomen of the GCA were recognized nationally and internationally for their extraordinary accomplishments during the Great War.[9]

Although the GCA continued to meet throughout World War I, the annual meeting of 1918 was canceled, presumably because of the outbreak of the Spanish flu. When the Executive Committee met in March, they "considered this matter most carefully" and "very reluctantly decided that the pleasant but unessential complications involved in such a meeting are better deferred to less troubling times." Although during World War II, GCA leaders would limit national meetings to those of the Executive Board, the only in-person annual meetings canceled during the entire history of GCA have been in 1918 and in 2020–2021, in both cases due to pandemics.[10]

Figure 2.7 The Woman's Land Army of America trained young women to take the place of men as farmers during World War I.

Reassessing Their Objectives

When peace came in 1919 and GCA members ended their war-related work, they reassessed their objectives and, after some debate, determined to stay their original course. From its first year of existence, the club had set up five national committees to appeal to its most urgent universal needs: 1) "to encourage the use of a color chart" in order to apply standards of color to garden plantings and flower arrangements; 2) "to inquire into opportunities for beautifying settlements and highways"; 3) "to inquire of seedsmen in regard to discount and purchasing in quantity"; 4) to seek out qualified speakers to give educational lectures; and 5) to assemble a collection of "garden literature," including not only books and papers but also botanical paintings and engravings. These national committees, which numbered seventeen by 1921, allowed the GCA to accomplish many of its broader objectives.[11]

At the annual meeting on Boston's North Shore in 1920, the GCA decided to incorporate; the organization now included forty-nine clubs. It was also in 1920 that Elizabeth Martin decided it was time to resign as president of GCA, having served the organization in that capacity for its first seven years. With her resignation, the GCA moved its offices from Philadelphia to New York City, where they remain to this day. Mrs. Martin continued to attend national meetings and would remain the honorary chair until her death in 1932. Not only would the GCA mourn her passing, but all

Figure 2.8 Elizabeth Martin encouraged garden club members to find peace and inspiration in their gardens on weekends and work for the war effort on weekdays. The garden of Lorelei Manning Gibson offers such a respite. (Lorelei Gibson)

of Philadelphia would: Flags were hung at half-mast there, honoring her as one of the greatest civic leaders in the city's history.[12]

While local garden clubs had always held flower exhibitions, in 1921 the GCA first participated in the International Flower Show in New York City, where the club displayed miniature garden models; the members again displayed such models in the 1923 show. Besides New York City, Boston and Philadelphia invited GCA to enter their flower and garden shows, and soon GCA members would participate in local, state, national, and international contests. The GCA prided itself on setting the standards for artistic floral and horticultural excellence as well as broadening knowledge of horticulture, flower arranging, and conservation.[13]

The National Arboretum: Patience and Persistence Rewarded

The advantage of being a national organization gave the GCA the strength to promote national issues about which they cared. One such cause, long held dear by GCA members, was the establishment of a national arboretum. The January 1914 *Bulletin* proposed, under the title "National Gardens," that a national arboretum and garden be established in Rock Creek Park in Washington, DC. Among its advantages, the selected property comprised one thousand available acres, making it a great site for "popularizing and reintroducing American trees and shrubs that formerly have been neglected" and for experimenting with new and valuable species. "These perfected plants should have their permanent place in our National Botanical Garden," the *Bulletin* article went on to proclaim.[14]

In April 1914 the *Bulletin* updated the GCA's progress in establishing a national arboretum and garden: "We are glad to be able to announce that the bill is now before Congress to establish a National Botanical Garden in Rock Creek Park, Washington D.C." As it turned out, however, the GCA's celebration proved premature. The bill's progress through Congress slowed to a halt, and it was not until 1920 that the legislators renewed their discussion of a possible national arboretum. Even then, apparently, congressional members' intentions were simply to implement improvements to the existing national garden, the United States Botanic Garden. This collection of botanic specimens, brought back from various locations, was housed in wood and glass greenhouses built in the 1860s at the foot of Capitol Hill in Washington, DC. When the GCA offered to address Congress on the

Figure 2.9 Beginning with the January 1914 *Bulletin*, GCA supported the idea of having a national arboretum. Congress finally approved the establishment of the US National Arboretum in 1934; it opened to the public in 1959. (Wolferman)

matter of a new arboretum, the lawmakers made it clear they did not want the club's advice. Undaunted, GCA members bided their time, waiting for a more favorable climate in which to advance their project.[15]

Some two years later, in 1922, the GCA Executive Committee approved the formation of a Washington Committee to work toward beautification of the nation's capital and establishing a national arboretum. Although GCA President Harriet Barnes Pratt did not think lobbying in Washington was a good idea for GCA, the chair of the Washington Committee, Janet Newbold Noyes, wife of Frank Noyes, owner of the *Washington Evening Star*, thought otherwise. Mrs. Noyes sought congressmen sympathetic to the idea of an arboretum and finally found a patron in Senator George Wharton Pepper of Philadelphia, brother-in-law of GCA founding member Rebecca Willing Pepper. Senator Pepper assumed leadership of the national arboretum cause, and in 1927, Congress finally approved the

National Arboretum Act. Dr. Frederick V. Coville, the first acting director of the National Arboretum, said the legislation "would not have been enacted at that time had not a committee of The Garden Club of America, under the chairmanship of Mrs. Frank B. Noyes . . . effectively urged its enactment."[16]

Janet Noyes and Harriet Pratt served on the first advisory board of the Council for the Development of the National Arboretum, which was enthusiastically supported by Secretary of Agriculture William M. Jardine. However, funding for the arboretum did not make the 1928 federal budget, and in 1929, when Arthur M. Hyde replaced Jardine as secretary of agriculture, it turned out that he did not share Jardine's enthusiasm for the project. One wonders whether Hyde, who had been governor of Missouri from 1921 to 1925 and had then resumed his law practice in Kansas City before President Herbert Hoover enlisted his help in his cabinet, might have listened more carefully to the GCA had The Westport Garden Club then been a member. However, Hyde cannot be blamed entirely for the arboretum's slow evolution, as the advance of the Great Depression soon overshadowed the new project. The GCA continued to monitor the progress of the arboretum, and, in 1934, Congress approved the establishment and funding of the United States National Arboretum in Washington, DC. GCA members voted to help with the development of the arboretum, which finally opened to the public in 1959. It would eventually become one of the country's largest arboretums and the only federally funded one. The GCA continues to support this research and education center, and GCA members have served continuously on the Arboretum's council.[17]

The Battle of Q37

In the 1920s, the GCA began what came to be called "the battle of Q37," a fight for a "reasonable quarantine law." The US Congress had enacted the Plant Quarantine Act in 1912, authorizing the US Department of Agriculture to inspect horticultural and agricultural products, to organize border quarantine inspections, and to restrict importation of infested agricultural goods. The GCA, naturally, had agreed that some kind of legislation was necessary to prevent foreign pests and diseases from entering the country. However, when, in 1927, Congress passed Quarantine Order 37, which further defined the Plant Quarantine Act, GCA members panned the

regulation changes as "draconian." Order 37 ruled out importation of many bulbs. It also required all imported plant materials, wherever their place of entry, to first be sent to Washington, DC, at the importer's expense, in order for the Department of Agriculture to examine and fumigate them. Delays in shipping and processing such a great number of plants in the nation's capital often led to their injury or destruction. GCA members, led by the avid Louise du Pont Crowninshield, began a full-fledged attack on the new quarantine rules. Herbert Hoover, then secretary of commerce, suggested to Mrs. Crowninshield that the GCA should go before Congress, prepared with an argument against American nurserymen who, because of the financial advantage it afforded them, supported Q37. However, the president of GCA, Harriet Pratt, countered that the GCA should not become an enemy of nurserymen. Instead, she and William A. Lockwood, the husband of GCA member Elizabeth Lockwood and the GCA's attorney for thirty years, testified before the Department of Agriculture concerning the disadvantages that Q37 engendered. The quarantine battle would continue, interrupted only when World War II curtailed importation of plants and spurred interest in propagating and hybridizing bulbs and other plants in the United States.[18]

After the war, Louise Crowninshield decided to take up the Q37 issue again. She visited congressional committees and voiced her opposition to the quarantine law, citing the backing, which was unauthorized, of eight thousand GCA members. By this time, the quarantine on imported plants seemed to have outlived its usefulness and the Department of Agriculture had grown tired of the burden of quarantines. Perhaps these reasons contributed to Mrs. Crowninshield's success, but, in any event, in July 1947 Congress enacted a new law that relaxed the quarantine on horticultural specimens and allowed bulbs to be shipped to the United States from the Netherlands.[19]

The Billboard Campaign

At the same time the GCA agitated for the repeal of Q37, another battle brewed, one that has continued into the twenty-first century. This issue had to do with the GCA's objective of beautifying the nation's roadways. One way to make highways and byways more attractive, the GCA said, when it established a billboard committee, would be to get rid of the "signboard menace." The issue of roadside signage and the prevalence of trash strewn along the sparsely paved roads and highways had bothered GCA members since the organization's founding. A paragraph in the July 1914 *Bulletin* read: "Our constant, unwavering ambition, is to utterly transform that no-man's-land of dishevelment and offense along the highways, those back yards, and those otherwise unoccupied limbos of cans and rubbish that mar our country and try our faith." In the GCA's 1920 minutes, all fifty-two clubs agreed that billboards were offensive. The nation's billboard lobby, however, was strong, and thus far taking the problem to court had not proved helpful. Essentially, the courts had ruled that a beautiful landscape was a luxury, not a necessity.[20]

In 1924 the GCA combined two committees to establish a formal Billboard and Roadside Committee, which continued to protest against road signs and to advocate for the preservation of wildflowers and trees. The committee urged "restriction of all outdoor advertising to commercial districts where it will not injure scenery, civic beauty, or residential values." When Edith Hope Goddard Iselin, from the North Country Garden Club on Long Island, became chair of the committee, a position she would hold for nine years, from 1922 to 1931, the club began to make progress on the billboard issue. In 1929, Mrs. Iselin announced that Massachusetts was spearheading a legislative battle against billboards. The members of all the GCA clubs in Massachusetts, with Mrs. Iselin's encouragement, attacked the billboard menace in every way their resourceful brains could devise. They gained the support of 258 national advertisers, who promised not to use highway advertising or, at least, not to renew contracts then in place. The success of the Massachusetts billboard campaign, however, was only the beginning. The campaign against billboards continued in other states, but with limited success.[21]

Advocating for Protection of Trees, Wildflowers, and Birds

In addition to the scenery-destroying roadside billboards, GCA members worried that urbanization and the growth of the highway system had led to the destruction of trees, wildflowers, and natural habitats for birds. GCA expanded its role in conservation during the 1920s and 1930s and encouraged local garden clubs to become more vocal in supporting local and state legislation that not only outlawed billboards but protected wildflowers, trees, and birds. When the GCA's former Wildflower Preservation Committee changed its name to the Conservation Committee in 1924, its chair, Fanny Day Farwell, called for the formation of three subcommittees: forestry, native plants, and bird protection, each with its own chair. Some members expressed concern that GCA was taking on too many causes. Martha Brookes Hutcheson, of the Short Hills Garden Club in New Jersey, rhetorically wondered "whether in finding billboards to suppress, rubbish to pick up, national parks to protect, Congress to influence, flowers to save, schoolchildren to inspire, and towns to plant, we are not in danger of losing sight of our original object, to set a standard (for) the finest gardens America can produce." Despite Hutcheson's concerns, the GCA's most visible early achievements clearly came in their conservation efforts.[22]

Saving the Redwoods

For many years the deforestation of redwood forests on the coast of California had disturbed GCA members. Early in the 1920s the GCA's *Bulletin* distributed printed pleas to save the redwoods from the voracious logging companies that were stripping thousands of acres of the majestic ancient trees each year. In 1929, Ethel Smith Landsdale, an active participant in the Save the Redwoods League, a nonprofit organization established in 1918, introduced an idea to make areas of public land off limits to logging: Groups or individuals could purchase tracts of redwood forest, with the State of California agreeing to match these private contributions. Landsdale, with her close friend Helen Stafford Thorne, a member of both the Pasadena Garden Club in California and the Millbrook Garden Club in New York, determined that GCA could foster this idea and become the vehicle for spreading sympathy across America for the plight of the redwood forests. The GCA members, tree huggers at heart, proved worthy adversaries to the loggers.

In 1930 a one-hundred-member delegation boarded the "GCA Special" train in New York City for the club's annual meeting, which convened that year in Seattle. After the spirited four-day meeting, the delegates, led by GCA President Elizabeth Lockwood and seventy-year-old Anne Thomas Stewart of Philadelphia, one of the founders of GCA and someone long interested in forestry, reboarded the "GCA Special" for Pasadena and the coastal redwood forests. Along with members of the Pasadena Garden Club, GCA delegates toured redwood groves, watched firsthand as loggers felled

thousand-year-old trees, and beheld hot dog stands and tourist concession shops set up amidst the noble trees. Most of these women were seeing the *Sequoia sempervirens* in person for the first time, a sight that awed them. Newly inspired, the women of the GCA took up the cause of saving the great trees with a vengeance. President Lockwood set up a Redwood Grove Selection Committee to oversee the establishment of a GCA-protected redwood forest and appointed Helen Thorne as the committee's treasurer. Thorne was dubbed the committee's "fairy godmother," and she became the project's largest single donor.[23]

At the GCA Annual Meeting of 1931, leadership encouraged the attendees to look beyond their local garden club efforts and support the national organization in saving the redwoods. By the winter of that year, the GCA had raised $91,634.13 toward its goal of $150,000, which would be matched by the State of California. Members from all over the nation, many of whom had never seen a redwood, and all ninety-four clubs in the GCA contributed to meet the goal and purchase a 2,567.72-acre redwood grove on the South Fork of the Eel River in Humboldt County, in northern California. The Garden Club of America Grove, the largest grove in the redwood forests of California, was dedicated in 1934. In 1943 the GCA Founders Fund Award allowed the organization to buy and protect fifteen acres of the Avenue of the Giants, also in Humboldt County. The next year Pauline Cropper Mallory of the Greenwich Garden Club in Connecticut suggested purchasing six thousand acres of redwoods on the northern coast of California in Del Norte County as a National Tribute Grove to honor US servicemen and -women. The GCA responded with gusto, contributing $52,000 to the purchase of this grove, dedicated in 1949. After the GCA Annual Meeting in 1952, 150 members visited the redwoods and reaffirmed that the club would provide everlasting protection for the National Tribute Grove and all remaining redwoods.[24]

Altogether, owing to the GCA's efforts to raise public awareness about their deforestation, seventeen groves and thousands of acres of coastal redwood forests have been preserved and still stand. Over the years, the GCA Grove

Figure 2.10 *Left:* Since the organization's founding, GCA has fought against billboards that disfigured roads and highways. Eliminating the "signboard menace" allowed for more beautiful vistas. (Kilroy)

Figure 2.11 In 1930 one hundred GCA members attended their annual meeting in Seattle and then traveled in their "GCA Special" train to tour redwood groves in California. Awed by the *Sequoia sempervirens* and angered by the loggers and hot dog stands there, GCA took up the cause of saving the redwoods and in May 1934 dedicated The Garden Club of America Grove in Humboldt County, California. (GCA Archives)

has expanded to 5,100 acres within Humboldt Redwoods State Park, and GCA has contributed more than $1.8 million to the conservation effort.[25]

Nevertheless, nothing can keep the redwood forests safe from disaster. In 2003, the Canoe Creek fire burned over eleven thousand acres in and near Humboldt Redwoods State Park, including some parts of the GCA Grove. Landslides and storms added to the damage done to the area's trees. While a great many of the redwoods survived and the ecosystem began to recover in the wake of the fire, many bridges and trails had been destroyed. Ten years after the disaster, in 2013 GCA launched a Bridge the Gap campaign to restore the River Trail, which had been left impassable. The trail reopened in 2018. All told, GCA members across the country had raised almost $400,000 to improve Humboldt Redwoods State Park.[26]

From August to October 2020, wildfires burned four million acres of California, including 16,000 acres of giant sequoia and 80,000 acres of coastal redwoods. (The 5,100-acre GCA Grove, located two hundred miles north of San Francisco, was not affected.) According to Save the Redwoods League (STRL), there is "no way to prepare in advance for the lightning wildfires that spread statewide" other than building fire resiliency with thinning and prescription burning. The GCA honored the dogged efforts of the members of the STRL in 2018 with the Elizabeth Craig Weaver Proctor Medal. The STRL's one hundred years of "unwavering dedication to and the protection and restoration of ancient coast redwood and sequoia forests" earned the respect of GCA, and STRL continues to warrant that trust. In 2022, the STRL acquired five square miles of privately owned coastline property in Mendocino County that was under direct threat of harvesting and development, thus saving what are known as the Lost Coast Redwoods.[27]

A 2020 study by Humboldt State University and the STRL determined that old-growth redwood forests "store more carbon per acre than any other forest type," confirming their key role in addressing climate change and the need for their preservation.[28]

In addition to their drive to save the redwoods, the GCA engaged in many other conservation efforts during the 1930s and 1940s. The organization acted to protect other endangered trees, including the holly, laurel, and ground pine. Bird sanctuaries, once considered a fad, now seemed essential to prevent extinction of some species. The organization supported a wide range of special projects and became involved in scores of conservation efforts.

Horticulture Plays a Major Role

In 1933, the GCA inaugurated a Horticulture Committee. Although horticultural issues had long interested GCA members, beginning with the horticulture report at the first meeting of the organization, now they would pay special attention to educating club members and the public about protecting and propagating native plants and promoting good garden design. The *Bulletin* began devoting an entire section of each edition to horticulture, and GCA flower shows, in addition to floral arrangements, displayed horticultural specimens. In 1937, Ethel Hallock du Pont, a member of the Horticulture Committee, started the Rare Plant Materials Project through which GCA members gave rare and practically unobtainable plants to nurserymen and arboretums for them to propagate and make available to the public. This practice would continue into the 1950s and 1960s with the Clematis Project, the Wildflower Project, the Daffodil Project, and others.

An early triumph for the Horticulture Committee was its display in the 1939–1940 New York World's Fair. Its "Gardens on Parade" exhibition occupied five acres with what the *New York Herald Tribune* called "the Most Stupendous, Most Magnificent, Most Gorgeous exhibition of flowers, shrubs, and horticultural beauties ever assembled." Harriet Pratt coordinated the GCA's efforts, enlisting the help of "several thousand individuals" from garden clubs all over the country, as well as many horticultural and botanical societies. Robert Moses, the head of the New York Park Board and the recipient of the GCA's 1938 Medal of Honor, proved to be a great friend and supporter of the "Gardens on Parade" exhibition. Conceived during the Great Depression and staged as Hitler invaded Poland in September 1939, causing France and Great Britain to declare war against Germany, the GCA's display at the World's Fair was remarkable.[29]

GCA's Contributions to the War Effort During World War II

The GCA's participation in the 1939–1940 World's Fair would be the club's last significant public display of beauty and tranquility before the United States entered World War II. In 1941, shortly after the attack on Pearl Harbor, Aline Kate Fox, the dynamic sixty-one-year-old president of GCA, received a telegram from the Defense Department with an invitation from Secretary of Agriculture Claude R. Wickard: "We shall much appreciate your attendance at the National Defense Garden Conference here December 1941 to assist in working out a National Defense Gardening program and coordinated plans for carrying such a program into action." Miss Fox immediately responded, "Of course we are attending," and named two GCA colleagues to accompany her. Kate Fox would, thus,

Figure 2.12 *Right:* Dedicated to protecting and propagating native plants, the GCA started its Horticulture Committee in 1933. The committee took on various projects, including the Clematis Project, the Wildflower Project, and the Daffodil Project. (Kilroy)

shuttle back and forth between her home in New York and Washington, DC, to oversee how GCA could help with the war effort. During the next forty months that the United States was at war, GCA clubs participated in projects such as building the landscapes of numerous army camps, planting and promoting Victory Gardens, and sponsoring lecture series on subjects ranging from nutrition and meatless menus to gardening. GCA also joined the Seeds for Britain Project in 1940 and by summer of 1942 had helped distribute American seeds to 1.2 million people in Britain, supplying 85 to 90 percent of the seeds for successful British gardens. All GCA member clubs participated in the war effort in some way and often collaborated with other groups like the Woman's Land Army and the Red Cross.[30]

Although garden club members took an active part in the nation's war effort, they also realized the importance of maintaining their own gardens. Miss Edith Kohlsaat, GCA chair of Zone 11 and a member of the Lake Geneva Garden Club in Wisconsin, expressed a common viewpoint: "We needed the peace and comfort of gardens [more] than ever before to keep us normal in these distressed times. . . . Garden clubs should retain their former identity, conserving beauty and doing everything along conservation lines." Beautiful gardens were deemed necessary for morale.[31]

During the war years, the annual GCA meetings focused on war-related business, but the club's first gathering after the war was a gala affair held in 1946 at the New Ocean House in Swampscott, twelve miles from Boston. Several garden clubs in Massachusetts—Chestnut Hill Garden Club, Cohasset Garden Club, Fox Hill Garden Club, the Garden Club of Buzzards Bay, Milton Garden Club, North Shore Garden Club of Massachusetts, and Worcester Garden Club—pooled their efforts to stage a GCA celebration for a whopping five hundred delegates. The festivities included two days of horticultural and conservation meetings. Tours of gardens, speakers, dinners, and business also occupied the delegates. In culmination, the Executive Committee proposed that the 1946 Founders Fund Project should entail the purchase of Bartholomew's Cobble, a site near Ashley Falls, Massachusetts, home to at least five hundred varieties of wildflowers. In making this land purchase, the GCA preserved yet another area of natural beauty. Delegates left the annual meeting anxious to go home and do their part to conserve wild plants, to promote horticultural education, and to beautify the many war memorials being dedicated around the country.

Figure 2.13 At the first GCA annual meeting held after World War II, the delegates agreed on a Founders Fund project to purchase Bartholomew's Cobble, a site near Ashley Falls, Massachusetts, that was home to at least five hundred varieties of wildflowers. (Virginia Bedford McCanse)

The Rise of the Environmental Movement

While horticulture remained at the heart of GCA objectives, the decade of the 1950s would prove to be the beginning of a national environmental movement. GCA already had a proud record on conservation, but now members became more aware of the damage being done to the environment and of the increased need to educate the public about ways to protect the nation's natural resources. In 1952, two members of GCA clubs in Connecticut, Edna Fisher Edgerton, a member of The Stamford Garden Club, and Anna Mark Rockefeller, a member of Horticulus in Greenwich, compiled a packet of information titled "The World Around You." This collection included six articles about our national resources and twelve leaflets on vital conservation issues. These packets proved to be very popular for use by classroom teachers and leaders of various organizations—so popular that the GCA issued tens of thousands of them over the next forty years. Ten years after the publication of the original study guides, the Conservation

Committee was still receiving almost seven thousand requests for them per year, coming from all fifty states and from some foreign countries. Because Mrs. Rockefeller kept the conservation materials up to date, "The World Around You" remained in distribution, through many editions, until 1992.[32]

One *Bulletin* article expressed the unique role The Garden Club of America thought women could and should play in the national conservation, preservation, and beautification effort: "GCA members, as women, have reason to know the lasting value of harmony in the home. Nature is the home of humanity; it is therefore the responsibility of women to do all in their power to instill, enhance and perpetuate harmony in this greatest of homes and families." Letter-writing campaigns remained the most often-used method of communicating the GCA's objectives, and at least sometimes government officials listened; but even when they did not, GCA members persevered. For example, the members of the Peachtree Garden Club in Atlanta had repeatedly suggested the need to beautify central Georgia rail stations by the planting of flowers. After club members sustained a long and vocal campaign, an official from the railroad contacted a GCA member, stating: "If you will call the Garden Club women off of me, the Central Georgia Road will plan and plant anything you want."[33]

GCA members had never shied away from making their views known to political leaders, and in the 1950s, GCA increased its advocacy for national legislation supporting its conservation goals. The GCA Conservation Committee had endorsed President Truman's policy establishing an airspace reservation over a portion of the Superior National Forest in Minnesota, which created a precedent for protecting wilderness areas.

When President Eisenhower took office, GCA conservationists wrote to him regarding the country's watershed policy: "We propose wise land use and the management of water run-off upstream to prevent erosion, diminish the danger of floods, and to facilitate the highest possible productivity of the land. . . . We further feel that the building of large dams downstream, flooding valuable land, and the Federal operation of power plants are unnecessary and encourage government extravagance, the growth of bureaucracy and socialism." Simultaneously, the Garden Club of St. Louis also took a stand against the proposed construction of large dams, which they argued did not prevent nor control flooding. President Eisenhower respected the GCA's input and even appointed a GCA Conservation Committee

member, Mrs. Halfdan Lee (formerly Mrs. William K. Jackson), as one of eighteen people to sit on the National Advisory Committee on Soil and Water Conservation under the direction of Secretary of Agriculture Ezra Taft Benson.[34]

In 1956, the members of The Westport Garden Club, eager to contribute to the national organization's continuing success in the areas of horticulture, preservation, and conservation, proudly joined The Garden Club of America. The Kansas Citians were well aware that the GCA's many achievements to date came largely through the efforts of individual member clubs and in turn through individual members. As the authors of a history of the GCA, *Fifty Blooming Years*, extrapolated, many of GCA's most important projects began with just "one crazy woman," a woman strong enough and determined enough to get an idea rolling and to gain the backing of her local constituents and of the entire GCA. It was a coup for both the GCA and The Westport Garden Club to affiliate, and, as it happened, WGC included more than one crazy woman who could and would make a difference in the landscape and environment of the United States.[35]

2.14 A bee finds nourishment on coneflowers. (Kilroy)

CHAPTER 3

The Early Years, 1956–1969

IN JUNE 1956, WHEN THE Garden Club of America invited The Westport Garden Club to join the national organization, WGC President H. E. Hockaday, "in a burst of extravagance," mailed handwritten cards to all Kansas City members announcing their new status. In red ink, she printed, "Flash: The Westport Garden Club has just been unanimously elected to membership in The Garden Club of America! Cheers!!" The WGC sent in its initiation fee and increased members' dues to cover the expense.[1]

Their affiliation with The Garden Club of America allowed WGC members to take advantage of the many opportunities offered by the national organization—to learn more about gardening and beautification efforts, engage in educational workshops, attend GCA conferences, participate in national flower shows, receive GCA publications, help influence state and national legislation, and take new forms of action to protect the environment. On the other hand, "overwhelming requests and directives arrived from headquarters and zone officers," according to member Lillian Marrs Diveley. WGC members responded enthusiastically, realizing that this was the time to move beyond gardening and creating flower arrangements to embrace the goals of The Garden Club of America.[2]

As GCA's 160th member, The Westport Garden Club became part of the vast and influential network that at the time was divided into nine zones, with WGC becoming part of Zone 8. (In 1969, GCA changed zone delineations

Figure 3.2 President H.E. McCune Hockaday sent postcards to all club members announcing that The Westport Garden Club had been admitted to The Garden Club of America.

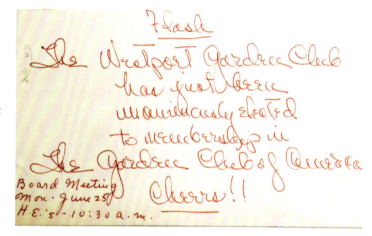

and increased the number of zones to twelve, also switching from Arabic to Roman numerals. The WGC then became part of Zone XI.) Each of the zones in GCA holds an annual or biannual meeting, and every spring GCA hosts an annual meeting to which each member club may send two delegates. In May 1957, the first WGC representatives attended the national group's annual meeting. Eight months beforehand, in September 1956, President Hockaday, Eda Luger, and Enid Kemper had attended the GCA zone meeting hosted by the St. Paul Garden Club in Minnesota. Hopefully, the three took good notes, because The Westport Garden Club would soon learn that the Lake Geneva Garden Club in Wisconsin had decided, for some reason lost in time, that it could not hold the next zone meeting. Instead, GCA asked the seedling WGC to host the meeting in October 1957.

Figure 3.1 *Left:* A burst of color characterizes many WGC members' gardens, including that of Blair Hyde. (Kilroy)

WGC Goes It Alone in 1957 Flower Festival

In spring 1957, before the zone meeting, The Westport Garden Club held its third flower festival at the Nelson-Atkins Museum of Art (which was then called the Nelson Gallery). Unlike the previous festivals, for this one the chairs—Katie Kessinger and Virginia Sartor Foresman—decided not to invite outside groups or individuals to participate. The show, once again, featured floral design entries in Kirkwood Hall, but for the first time, the museum opened the loan gallery to the WGC. Fourteen members used the space to set tables there, using their own silver and china accompanied by outstanding floral arrangements. *The Independent* enticed festival-goers not to miss the table settings, promising that "rare porcelain and antique silver from private collections, never before shown, will be effectively displayed at the exotic Festival" certain to showcase "the ultimate in floral beauty." The net profit from the festival amounted to $1,804.91, with the museum receiving $900. Museum Director Laurence Sickman noted in his annual report that "the generous cooperation of the members [of The Westport Garden Club] is a valuable asset."[3]

Figure 3.3 The WGC sent invitations to the 1957 Flower Festival.

Figure 3.4 *Left:* Lilies bloom in Blair Hyde's garden. (Kilroy)

WGC Hits a Home Run

The Westport Garden Club had little time to recover from the third flower festival before preparing to host the zone meeting, the first GCA event ever held in Kansas City. Zone meetings require a great deal of preparation time, which the WGC did not have. However, honored to have been asked to host the meeting, club members enthusiastically, if a bit frantically, rose to the task. Although they typically had most of July (after their first Monday meeting) and all of August to rest before their meetings started up again in September, July and August of 1957 proved anything but relaxing. H.E. Hockaday and Virginia Foresman, who had agreed to co-chair the zone meeting, put fifteen planning committees in place to figure out how best to house, transport, entertain, and feed all the visiting delegates while showing off members' gardens and the civic projects the young club already had spearheaded. Mrs. Hockaday informed club members that they would be responsible for housing the sixty-one guests and hosting private dinners in their homes on the first night of the convention. The September membership meeting reviewed the zone meeting schedule. Shortly thereafter, a luncheon at Patti Abernathy's home convened at which committee chairs fine-tuned and cemented the details of the zone meeting lined up for October 1 and 2.[4]

Besides the sixty-one delegates who represented the GCA clubs in Zone 8, Zone Chair Mrs. Harold Seaman of Milwaukee and GCA President Mrs. Randolph C. Harrison of Greenwich, Connecticut, attended the meeting. Mrs. J. Churchill Owen, of Denver, who had come to Kansas City two years earlier to judge the 1955 Flower Festival, was the guest of her longtime friend Helen Beals. The two-day event began with a business meeting at the Nelson-Atkins, followed by a tour of the museum and lunch in Rozzelle Court, the enhancement of which WGC members were proud to point out as one of their successful civic projects. After lunch, Director Sickman spoke, followed by GCA President Harrison. That afternoon the delegates visited Rosehill Nursery and Garden and the Linda Hall Library, billed by Mrs. Seaman as "a depository for all atomic energy publications." The Zone 8 delegates were suitably impressed by the "library's luxurious assembly room."[5]

On Wednesday, October 2, attendees convened at the Carriage Club and participated in roundtable discussions on conservation. After lunch,

Figure 3.5 WGC President Virginia Page Hart and Julia Jackman Bartlett, whose husband was chairman of the Linda Hall Library trustees, in front of the lead fountain with the Mercury figure that the club gave to the library. (*The Independent*, September 21, 1957)

Figure 3.6 Zone delegates were impressed by the beauty of the trees in Kansas City. (*Trees*, by Margot Munger Peet)

they heard presentations on the GCA Founders Fund, horticulture, and national parks. At 2 p.m. delegates embarked on a bus tour of six gardens, followed by cocktails on Helen Houston and Richard Nelson's patio and culminating in dinner at The Kansas City Country Club. After dinner, Charles van Ravenswaay, an author, horticulturist, and historian dedicated to the preservation of Missouri's architectural history, spoke about historic sites in the Midwest, thus concluding the zone meeting.

Apparently, despite the members' having had little time to prepare for the meeting, it was an overwhelming success. Zone Chair Mrs. Seaman wrote, "The Westport Garden Club can be very proud of their achievement. It was a charming meeting with every detail well planned. Mrs. Hockaday, Mrs. Foresman, Mrs. Hart, Mrs. Townley, and the members of their committees should be congratulated on the wonderful work they did." Mrs. Seaman was also "greatly impressed with the beauty of the trees and the interesting topography of the land," a common revelation to people who have never visited Kansas City.[6]

Developing Expertise in Floral Design: Husbands Get in the Act

Amid hosting a zone meeting and putting on flower festivals, the WGC continued furnishing floral arrangements for museum events. In fact, Debbie Thompson Gates (the granddaughter of member Elizabeth "Betty" Searle Thompson and daughter of member Prudence Townley Thompson but not a member herself) equated the garden club solely with flower arranging, which was not a complete overstatement. The museum called on the WGC ladies to decorate for all Friends of Art shows and meetings and all manner of other occasions. When the American National Red Cross Association hosted four thousand delegates at the museum in 1960, Helen Thompson took charge of coordinating the flowers. Before the exhibit, *The Imagination of Primitive Man*, opened in 1962, Harriet "Hattie" Collins Byers and a team of ladies created a special arrangement around a "house post," as Curator Ted Coe referred to the artwork. For the *Art Treasures of Turkey* exhibit in 1967, Florence Boyd Beaham, Barbara Forrester Rahm, and Hattie Byers decorated the tables.[7]

Figure 3.7 WGC members furnished beautiful floral arrangements for museum events. (Bouquet by Martha Lally Platt)

Floral design was so key to the success of The Westport Garden Club that the club brought in knowledgeable authorities to provide new ideas for members. In 1956 Mrs. Gilbert Miller, whose photographed arrangements appeared in *The Kansas City Star*, presented a program at Eda Luger's home. Her technique involved placing plant material firmly and stiffly in place in a stylized manner. In contrast, in 1958, Shozo Sato, a Japanese flower arranger who in 1965 would write the definitive book on ikebana (*The Art of Arranging Flowers: A Complete Guide to Japanese Ikebana*), held a class at the Nelson-Atkins. Sato demonstrated how to bend a pussy willow branch and to balance three narcissus in a tall, glass vase without a holder. In September of the same year, Georgiana Reynolds Smith, of Dover, Massachusetts, gave a two-day lecture and workshop entitled "Creative Flower Arrangements Related to the Fine Arts." The club brought Elizabeth Reynolds to Kansas City in May 1964 to present a three-day workshop on flower arranging, and in May 1967 Julia S. Berrall, author of *A History of Flower Arrangement*

(published in 1953 and reissued in 2000), conducted a three-day workshop. Much like clothing fashions, flower-arranging techniques had a broad range and the popularity of certain flowers and styles changed with the times, but members of The Westport Garden Club always did their best to keep up with current trends.[8]

Having forgone plenty of dinners because their wives were arranging flowers or doing garden club business, the WGC's "men's auxiliary" decided one Monday in the late 1960s to try their hand at flower arranging. The men met secretly at Bob Legg's florist shop, where they created centerpieces for the WGC dinner to be held at The Kansas City Country Club the following evening. When the ladies arrived, one grande dame, perhaps Helen Delano Sutherland, exclaimed, "Who did these awful centerpieces?," completely unaware that her husband was one of the offending "florists." The criticism was duly registered, if not too seriously, for those "awful centerpieces" would not be the last arrangements the men made for club events.

In due time, the men's auxiliary exacted a bit of revenge. In 1977, they actually came to a club dinner dressed as "lovely ladies." They proceeded to critique the centerpieces that their wives had assembled, using GCA vocabulary they had pilfered from The Garden Club of America's *Flower Show and Judging Guide*. Continuing the repartee, in 1981 WGC members asked interested spouses to make floral arrangements for another club dinner; this time the women judged them, using GCA standards. First prize went to Jack O'Hara, with the following comment from the judges: "Relates well to container. Good composition and interesting interplay of materials." Charles Seidlitz garnered second place because of his "splendid distribution of color and textures ranging from plastic apples to seashells and pomegranates," while third place went to C. Humbert Tinsman and honorable mention to Herman Sutherland.[9]

Some twenty years later, the WGC ladies again recruited the men's auxiliary to arrange flowers for a dinner. While the women held a business meeting in a separate room, they turned the men loose with buckets of flowers and vases to make floral arrangements for the dinner tables. When the women came out of their meeting, they realized that no one had bothered to tell the men they needed to sign their own tickets at the bar. The bar bill, which was truly awesome, was paid by the WGC. No mention was made of the bouquets.

The men's auxiliary had another opportunity to arrange flowers at the WGC Christmas party in 2019. This time there was no confusion about drink tickets. The WGC ladies socialized while Alison Wiedeman Ward divided the men into teams and directed them to create "arrangements" using "everything" in a bucket of flowers and ribbons. One floral design was memorable for its title: "Eat Your Heart Out, Martha Stewart," but the winning arrangement was one that used *everything* in the bucket, including the bucket itself. The creativity of the men's auxiliary lives on!

Award-Winning Designs

As soon as the WGC joined the GCA, the ladies from Kansas City began entering GCA flower shows and, in fact, winning awards. Seventeen GCA clubs participated in the 1962 Chicago World Flower and Garden Show, at which three WGC members won awards. Marjorie Paxton won second prize for "an arrangement of vegetables in pewter," her arrangement deemed an "amusing, gay, and interesting conception. The romaine and broccoli were well handled. The artichoke too dominant. Nice variation of textures." Lillian Diveley won third prize with her arrangement, while "The Ivy Eagle," the entry of Flora Markey Barton, took first place as well as the Award of Merit. Years later, reminiscing about his grandmother, Bruce Barton recalled how much she loved her roses and how devoted she was to the WGC. Grandmother Flora could not have been better named![10]

Other flower show awards in these early years went to Thelma Williams Frick, who won a silver medal in the Massachusetts Horticulture Society Show in 1964. The Westport Garden Club won a blue ribbon in 1965 in the American Flower and Garden Show of Kansas City, where Barbara Rahm won a red ribbon. She went on to win a blue ribbon at a show in Colorado Springs in 1968.

Figure 3.9 WGC member Thelma Williams Frick's garden in Cotuit, Massachusetts, won a silver medal in the Massachusetts Horticulture Society Show in 1964. (WGCA)

Figure 3.8 Helen Delano Sutherland poses with Flora Markey Barton to admire her "Ivy Eagle," which took first place in the 1962 Chicago World Flower and Garden Show. (*The Independent*, June 16, 1962)

Christmas Traditions and Revivals

In December 1957, after hosting the zone meeting, WGC members enjoyed a festive holiday luncheon at The Kansas City Country Club. Club members made plans to spread the holiday spirit by bringing jars of "gaily wrapped jams and jellies" to distribute to "convalescent homes." This tradition would endure for more than thirty years until, finally, the nursing homes said, "Enough!" Even today "gaily wrapped jams and jellies" is a refrain recited with humor by all veteran members.[11]

The 1957 holiday luncheon also included the first "Legends of Christmas" pageant, a presentation written by Lillian Diveley. After extensively researching holiday symbols and traditions of many lands and writing a script, Mrs. Diveley then strongly encouraged creative club members to help assemble

A Meeting of the Westport Garden Club
will be held at

Kansas City Country Club

Mrs. Philip F. Rahm, Luncheon Hostess

Date Tuesday, Dec. 3rd Time 10:30 A.M.

Please bring two small
jars of jelly, individu-
ally gift wrapped.
R.S.V.P.

Vivian Foster
Corresponding Secretary

Figure 3.10 Vivian Pew Foster sent out typed invitations to the club's December 1957 meeting.

Figure 3.11 A table at The Kansas City Country Club was set up to receive the "gaily wrapped jams and jellies." (WGCA)

an exhibit of Yule-time decorations to illustrate her text. Gregorian music played in the background while Mrs. Diveley read her narrative.

Since the "Legends of Christmas" pageant proved to be very labor-intensive, members decided to reenact it only every five years. Thusly, the program was repeated in 1962, 1967, 1972, 1977, 1982, and 1987, when, according to *The Independent*, Mrs. Diveley "again this year" recounted "one enchanting story after another." A symbolic decoration created by an artistic club member illustrated each story, and a "lamplighter," who was always a new member, lit a candle as the stories unfolded. DeSaix "DeeDee" Willson Adams remembers being the lamplighter in 1987, having joined the club in 1985. She also recalls that Lillian Diveley had asked her to create a children's table as part of the Christmas pageant, which DeeDee did with some trepidation, as Mrs. Diveley was a bit intimidating and no one could say no to her.[12]

Lillian Marrs Diveley: A Penchant for Perfection

If one mentions the name Lillian Diveley to longtime garden club members, one might perceive a frisson of apprehension in them, somewhat akin to being called to the principal's office. Lillian, a founding member who served as president of the club from 1954 to 1955, is remembered for the exacting perfectionism she demanded, both of herself and others, at times in a rather forceful manner. That said, it is probably due to her influence that The Westport Garden Club gained its reputation for hosting events with impeccable attention to detail and in exquisite taste.

Lillian was married to Dr. Rex Diveley, a prominent Kansas City orthopedic surgeon, whom she always referred to as "The Doctor." The dapper Dr. Diveley, a skilled photographer, was often recruited by Lillian to record club events and was a popular member of the so-called men's auxiliary.

Always immaculately turned out, Lillian had a background in fashion, having worked as a designer for the Donnelly Garment Company, later called Nelly Don. Her skills were put to use whenever the club needed tablecloths or fabric bags for fundraisers or zone meetings. Lillian's favorite color was yellow, and when she "suggested" yellow as the theme color for the 1987 zone meeting, no one dared disagree.

Figure 3.12 "The Doctor" Rex and Lillian Marrs Diveley enjoyed time together in their garden. (WGCA)

Lillian was an approved GCA floral design judge, and in 1989, shortly before her death, she received the GCA Club Medal of Merit for her contributions to the WGC and to her community. Not usually one to display emotion, Lillian was moved to tears by the honor. ✿

After Lillian Diveley's death in 1989, members would from time to time resurrect simulations of the "Legends of Christmas" pageant. One year Laura Kemper Toll Carkener wrote a program she called "The Plants of Christmas," which featured greens and flowers used in pagan festivals as well as by the Romans, Germanic tribes, Christians, and Druids. In 1995, Elizabeth "Betty" Kennedy Goodwin planned a new Christmas program called "Traditions of Christmas" to honor Lillian Diveley, as it was she who had proposed Betty Goodwin for membership in the WGC. Again, Gregorian chants provided background music, and candles were lit as each vignette unfolded. In a change of pace, the 1995 program featured personal family traditions narrated by members who wanted to participate. Seventeen years later, Eulalie "Eulie" Bartlett Zimmer suggested a "Traditions of Christmas" revival, and on December 4, 2012, the "soft sound of Gregorian chants filled the darkened room" at The Kansas City Country Club once again. Members illustrated their family holiday traditions with beautiful presentations. Cynthia "Cindy" Rapelye Cowherd remembers sharing a collection of miniature, hand-painted wooden figures and ornaments from the Erzgebirge region of Germany that had belonged to her mother, WGC member Sarah "Sally" Rapelye Cowherd.[13]

The Fourth and "Most Ambitious" Flower Festival: 1959

Not long into 1958, the club began preliminary plans for their most ambitious flower festival yet, to be held the following year. The first three flower festivals, held three years in a row, were, as Katie Kessinger explained, "highly experimental" and not nearly as elaborate as those that followed. Inspired by the Rittenhouse Square Festival held in Philadelphia, the 1959 event and subsequent flower festivals, in 1961, 1963, 1966, and 1969, required a great deal of effort from all fifty active members; nonetheless, the club deemed the successful results validation of the undertaking.[14]

Figure 3.13 The club held its most ambitious flower festival in 1959.

Josephine "Jo" Reid Stubbs, who would serve as president of The Westport Garden Club from 1959 to 1961 (the first to serve a two-year term), chaired the fourth flower festival with Co-Chair Laura Kemper Carkener. The club had decided that the proceeds from that year's flower show would benefit the creation of a new, improved junior art center at the Nelson-Atkins Museum of Art. With very high expectations and more than a year to plan, much effort and many innovations went into preparing for the April 1959 festival, themed "Flowers Around the World."

Of course, as with all events held at the Nelson-Atkins, everything had to be preapproved by the museum's trustees, who continued to express their concern that the museum was an art museum whose primary mission was to educate the public about art. The trustees' minutes for September 30, 1958, note that much discussion ensued about the appropriateness of including a fashion show in the flower festival, something the ladies determined would enhance attendance. Whether or not the trustees approved this new development, it happened anyway, with programs at 2:00 and 7:30, where models showed "fashions for garden parties and garden digging," as *The Kansas City Star* put it. In photographs from the event, the only real distinction between the two kinds of garb seem to be that a frilly, flower-bedecked hat was essential for a garden party, while a more protective head covering was desirable for gardening.[15]

Figure 3.14 Club members modeled fashions for gardening and garden parties at the flower festival. (*Kansas City Star*, April 23, 1959)

FASHIONS FOR GARDEN PARTIES and garden digging will be featured in style shows at 2 and 7:30 o'clock tomorrow at the fourth annual Flower Festival to be sponsored by the Westport Garden club at the Nelson Gallery. Mrs. George P. Luger (left) wears a bright red patterned blouse and blue denim skirt for gardening. The rooster accent on the skirt is repeated in the blouse material. Mrs. L. Patton Kline (right) models an embroidered white organdy dress with a picture hat for a garden party. Thirteen of the 28 models will be daughters and daughters-in-law of club members. The festival will be held from 12 until 5 o'clock and from 7 until 9 o'clock. Proceeds will be used to establish a fund for a children's Junior Gallery.

The Fashion Show Committee was just one of six committees that put on the 1959 WGC Flower Festival. The others were charged with overseeing floral display, artists' booths, children's art exhibitions, door prizes, and the garden and food market. While the center of Kirkwood Hall was filled with floral arrangements, booths lined the hall's marble walls, offering for sale various foodstuffs as well as house plants, flowering pots of annuals and perennials, vegetable seedlings, and ivy trees—most of which had been grown by club members. Again, the loan gallery housed the table settings provided by WGC members.

On the day of the big event, WGC members donned cherry red broadcloth smocks, made by seamstress Irene Zagar for five dollars per smock, which did not include the cost of the materials. These smocks became a tradition for many years, as did wearing plastic dogwood pins, the dogwood, of course, being the club's symbol. The pins further distinguished members from attendees of the show. In 1973, needlepoint dogwood flower pins replaced the plastic pins. Artist Frederic James, husband of WGC member Diana Hearne James, designed the logo, while Joan Knight Sherman, daughter of WGC early member Martha Knight, painted the needlepoint design onto canvases, which she sold from her needlepoint shop. Although at the advent of the needlepoint pin, members needed to stitch them for themselves or hire someone to do so, eventually it became an unwritten (though not always followed) tradition for a sponsor to needlepoint a pin for a new member she proposed. Frequently, grandmothers or mothers passed their pins on to daughters or granddaughters, relieving the proposer of the responsibility.

Figure 3.15 Red broadcloth smocks made their first appearance in 1959 and continued for decades to identify WGC members at flower festivals. From left: Margaret Hall, Betty Berry McLaughlin, Margaret "Peggy" Garner Gustin, Suzanne "Susie" Slaughter Vawter. (WGCA)

Figure 3.16 Needlepoint dogwood pins replaced the plastic variety.

It was also at the 1959 festival that, for the first time, daughters and daughters-in-law of WGC members helped out in the booths, providing a much-needed boost to the limited womanpower of the club. The volunteers donned corsages, not yet having earned dogwood-pin status.

The proceeds from the festival totaled $2,350.89, outdoing all previous festivals, but, according to Katie Kessinger, it had "not [amounted to] what we felt necessary to give the help needed for the new junior gallery, so we augmented it in order to make a gift of $5000."[16]

WGC Plays a Major Role in Art Education

Since its inception in December 1933, the Nelson-Atkins Museum of Art had considered education of children a top priority and had offered several children's art classes. Museum Director Paul Gardner also wanted to offer guided tours of the art collection to area schoolchildren, but the museum did not have the staff or the funds to do so. However, at a dinner party held the winter after the museum opened, Gardner serendipitously found himself seated next to Jane Hemingway Gordon, the chair of the Arts Committee of the Kansas City, Missouri, Junior League and a future WGC founding member. Mrs. Gordon suggested to Director Gardner an idea that they quickly implemented: that Junior League members, several of whom had served as hostesses at the museum opening, would be interested and could be trained to give tours of the museum. Beginning in 1934, members of the Junior League began conducting museum tours for public school sixth graders in Kansas City, Missouri. In that very first year, twenty-four Junior League members gave tours to 23,200 children. The museum expanded the tour program in the 1940s to include students in elementary grades in both private and public schools in the area. By 1960, 1,104,517 students had taken museum tours. The volunteer docent program Mrs. Gordon had started was being mimicked by the Metropolitan Museum of Art in New York as well as by countless other art museums.[17]

In the 1940s the Nelson-Atkins decided to broaden the scope of children's classes and to enhance the area used to display the children's artwork. In a unique twist, the exhibits in the corridor outside the classroom area were planned for children by children, and a junior docent program was established whereby older children guided younger ones through what came to be called the "Little Museum." When James E. Seidelman became director of the museum's Department of Education in 1952, the art education program expanded further. According to the museum's former director of education, Ann Brubaker, who wrote the history of junior education at the museum, the Creative Arts Center under Seidelman became "a nationally recognized model for museum-based education programs for children." When the trustees and the museum's director expressed the need for a proper area to display children's art and a place to further children's art education through visual aids, the WGC came to the rescue. The club agreed to sponsor an enlarged Junior Gallery and Creative Arts Center, "with the stipulation that the educational staff set up a program and teach horticulture and conservation to the children."[18]

The Westport Garden Club grant of $5,000 enabled the remodeling and expansion of the Junior Gallery, which would continue the traditions of the

Little Museum while providing more extensive exhibition space to display children's artwork. The opposite side of the corridor was equipped with audiovisual aids. In addition, WGC funds provided for the conversion of a former classroom into a "Gallery for Interpretive Exhibits which teach the children the historical methods, materials, and techniques used in the various arts media, thus giving them a rich understanding of the masterpieces of the Nelson Gallery of Art."[19]

Figure 3.17 Children enjoyed the opportunity to show off their artwork at the Little Museum at the Nelson-Atkins Museum of Art. (WGCA)

In January 1960 the enlarged and remodeled Junior Gallery and Creative Arts Center opened to much acclaim. The redesigned corridor and classroom represented one of just four major children's galleries and centers in the nation. The trustees of the museum, along with The Westport Garden Club, invited art educators, school personnel, and local leaders in education to a tea and a preview of the Junior Gallery on Saturday, January 30, 1960,

from 3 to 5 p.m. With the dedication of the Junior Gallery, the trustees also took the opportunity to celebrate the museum's Education Department and the Junior League docent program, both of which had been started in 1934, more than twenty-five years earlier.[20]

The previous evening, January 29th, at a black-tie dinner for three hundred people given by the museum's trustees, the members of the Junior League of Kansas City, Missouri, and The Westport Garden Club, the Junior League announced their gift of a new Junior Library: "This gift not only commemorates twenty-five years of Junior League service to the children of the Kansas City area, but dedicates a continuing interest in the cultural development of our future citizens." Simultaneously, The Westport Garden Club revealed a plaque with an inscription written by Director Sickman to be hung outside the new Junior Gallery.[21]

Figure 3.18 Museum Director Laurence Sickman chose the words for the Junior Gallery plaque, 1960. (WGC Records, RG 71/01, NAMAA)

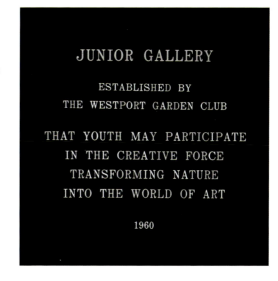

Louise Conduit, the supervisor of the Metropolitan's Junior Museum, spoke at the dinner on the importance of children having a museum of their own. "Schools cannot teach children enough about art," she said, "as there are already too many pressures on schools." When Conduit had responded to her invitation from the trustees to speak at the Junior Gallery opening, she told them that she had followed the Nelson-Atkins program of education for twenty-three of its twenty-five years "with admiration and appreciation." At the top of her invitation, she wrote this quote, which she used in her speech: "We are children once in a lifetime of art. The way we experience it in our youth may open its world of seeing, enjoying, and creating for the rest of our lives."[22]

While the display area of the Junior Gallery was a substantial improvement over the previous area allotted for the Little Museum, the audiovisual area, in the corridor, drew the most attention. Six white telephones lined the hallway, each with a corresponding peephole. When a visitor lifted one of the receivers, a young voice said "Hello" and then explained what was behind the peephole. Some of the peepholes, situated at a child's eye level, showed color slides of paintings; others had filmstrips, which the viewer could manipulate by turning a dial. A *Kansas City Star* reporter who saw this area in action remarked, "To the children, it is like one big package of surprises. . . . Telephones and peepholes intrigue children and excite their interest in art." As the director of the Junior Gallery and Creative Arts Center, James Seidelman, stated, "I speak for the youth of greater Kansas City in expressing appreciation for the dedicated interest and support of The Westport Garden Club." In an article Seidelman wrote for *Museum News* in April 1961, he related that the museum's dream of having an educational program for children from three to sixteen years old was coming true, a dream that had begun with The Westport Garden Club's initial grant of $5,000. "Their interest in the educational program," Seidelman continued, "stems from the thought that it is only with the guardianship of the youth of today that nature, beauty, and art can be preserved."[23]

The Westport Garden Club's end-of-the-fiscal-year luncheon was held in June 1960 at the Loose Park Garden Center. The meeting celebrated the culmination of a very special year that included the dedication of the Junior Gallery and Creative Arts Center. No one was more excited about the opening of the new children's facilities at the Nelson-Atkins than former Kansas Citian Katherine Harvey, daughter of Ford Harvey and granddaughter of Fred Harvey, the man who had "civilized the West" by developing the Harvey House restaurants, souvenir shops, and hotels, serving passengers on the Atchison, Topeka, and Santa Fe Railway. Katherine Harvey had moved to Pasadena, California, where she became a member of GCA's Pasadena Garden Club. However, she would always have a warm affection for her hometown and for the Nelson-Atkins. She had even hosted a dinner party the night before the museum's opening in 1933 for dignitaries and

Figure 3.19 Children learned about art by using the Junior Gallery phones, 1960. (Department of Education Records, RG 32, NAMAA)

out-of-town guests, including art dealers from New York, London, Paris, and Venice. Therefore, to congratulate the WGC ladies on the impact they had made at the museum, Katherine Harvey shipped nine cymbidium orchids from her Pasadena gardens to decorate the Loose Park Garden Center for the June meeting.[24]

An interesting aside is that Katherine Harvey's goddaughter Blair Peppard Hyde, a member of The Westport Garden Club since 2000, had been named Blair by her parents at Katherine Harvey's request: Blair was Katherine's mother's maiden name. Blair Peppard Hyde, a past president of the WGC, is the daughter of founding member Marcella Ryan Peppard and the niece of founding member Ann Peppard White.[25]

In 1964, The Garden Club of America awarded The Westport Garden Club the Amy Angell Collier Montague Medal for outstanding civic achievement, commending the club "for its vision, its energetic and continuing advisory, manual and financial help in making the Junior Gallery

Figure 3.20 Katherine Harvey shipped cymbidium orchids from her Pasadena garden to congratulate the WGC on the installation of the Junior Gallery.

Figure 3.21 The Garden Club of America awarded The Westport Garden Club the Amy Angell Collier Montague Medal for outstanding civic achievement in 1964. (WGC Records, RG 71/01, NAMAA)

and Creative Arts Center of The Nelson Gallery and Mary Atkins Museum of Art in Kansas City, Missouri, an important part of the cultural growth of the community." The award was presented in Portland, Oregon, at the 51st annual GCA meeting. Martha Knight, as WGC president, accepted the medal for the club. With her were WGC members Patricia "Patty" Castle White and Ruth Carr Patton Kline, who was also a director on the GCA's national board. *The Kansas City Times* headlined the WGC's achievement, reporting that the club had won a Garden Club of America medal for establishing the Junior Gallery, which, the article said, "offers children a unique program of relating art and nature studied together in the classroom and in walks on the extensive gallery grounds."[26]

Transforming Nature into the World of Art

A core objective of GCA is education, and, while teaching others about horticulture had long been a purpose of the national organization, its efforts in conservation education had lagged. By supporting children's nature/art classes at the museum, The Westport Garden Club would make a unique contribution toward furthering the GCA's goals. Classes in the new Creative Arts Center combined art with horticulture and conservation, as requested by the WGC. Projects varied by age group. Even very young students loved finding and then drawing friendly garden dwellers, including toads, earthworms, bees, and wrens. Kindergarteners made birdseed gardens, while fourth graders collected leaves on the museum grounds to use in creating ink rubbings, silhouettes, and spatter prints. In a ten-week course, older students studied the museum's stands of native trees, which included pin oak, scarlet oak, hackberry, sugar maple, red maple, sweet gum, beech, and birch trees, and made dioramas that illustrated how water, soil, and forest conservation and maintenance contributed to a beautiful city. Another class, entitled Discovering Nature, Beauty, and Art, called upon each student to adopt a tree, sketch it, study it, and learn how one might protect it. The Green Thumb Series emphasized an artistic approach to gardening in a six-week program held from 10 a.m. to noon on Saturdays for children ages six to sixteen. These kinds of classes continued for years.[27]

The Creative Arts Center's nature program extended beyond the museum walls. In the spring of 1961, children from the museum's art classes took an Art-Nature Tour, which included the gardens of five WGC members: Millie

Hoover, Helen Nelson, Dorothy Dillon Cunningham, Maxine Goodwin, and Lillian Diveley. For several years, art-class students took trips to WGC members' gardens, where they made sketches of what they saw. Then, back in the studio, they translated their sketches into finished artworks for display in the Junior Gallery.[28]

Figure 3.22 Dorothy Dillon Cunningham invited children from the museum's art classes into her garden to make sketches. (WGCA)

Gardens Welcome Tours

Children enrolled in classes held at the Nelson-Atkins were not the only locals privy to visiting WGC members' gardens. Millie Hoover welcomed visitors to her gardens at 5651 High Drive for the fourth annual garden tour of the Garden Guild of St. Andrew's Church in 1956. In June 1958, WGC President Julia Tinsman showed off the garden she was developing at her French country home on Verona Road, which was inspired by gardens she and her husband had visited the year before in Burgundy, France. In June 1961, WGC members celebrated the end of the club season in the "bright and colorful garden of Dr. and Mrs. John Swann Knight, afire with arbors of Paul Scarlet roses," as reported in *The Independent*.[29]

WGC members and their husbands enjoyed a very special garden tour in June 1962 in honor of the men's auxiliary and to celebrate Father's Day. The group toured Linda Hall Library and gardens at the homes of founding members Jane Gordon, Margot Peet, Laura Kemper Carkener, Lillian Diveley, and Maxine Goodwin. The Goodwins served cocktails in their garden before all the members and guests went to The Kansas City Country Club for dinner, where the main topic of discussion was, according to *The Independent*, "who has the greenest thumb."[30]

Flower Festivals of the 1960s

The junior education center's success inspired such enthusiasm among both the museum staff and WGC members that the club decided that the fifth flower festival, to be held April 28, 1961, would have children as its theme and would again benefit the Junior Gallery. Margaret "Peggy Sue" Neal either volunteered or was recruited to be the festival chair. Peggy Sue was one of the grande dames of Kansas City, and her home on Brookbank Lane, just down the street from the Goodwin house, was a gathering place for good friends. Her style as an interior decorator was well known, but she was also active in various civic activities as a member of the Junior League of Kansas City, Missouri, past president of the Smith College Club, and trustee of The Barstow School.

Whatever her "volunteer" status, Peggy Sue Neal carried out her role as festival chair with panache. For the show, the WGC placed a carousel, eighteen feet in diameter, in the center of Kirkwood Hall, and after the show Peggy Sue made sure that the museum converted the walls of the carousel, which had cost $250 each, into display boards for the Junior Gallery and Creative Arts Center. During the show, WGC members and their auxiliary staffers sold works of art made by children from both the Kansas City, Missouri, public schools and the Junior Gallery classes. Director of Education Seidelman selected six little girls as part of the sales staff for the children's artwork booths. Besides the children's section, local artists displayed their work for sale. The festival also included a garden booth with live plants as well as garden furniture and accessories, which club members set up in the south vestibule. WGC members added a merchandise mart of consigned items along with the first Collector's Corner, which offered donated items for sale. Although the 1961 festival was geared toward children, the flower arrangements were inspired by the *George Caleb Bingham Sesquicentennial Exhibition*, which had opened at the museum on March 16.[31]

Figure 3.23 Peggy Sue Neal chaired the 1961 Flower Festival, which again benefited the Junior Gallery. She had a huge carousel erected in Kirkwood Hall, the walls of which would later be converted into display boards in the Junior Gallery. (*Kansas City Star*, WGCA)

The Independent had predicted that the 1961 festival would "outdistance preceding editions in riotous color, splashed in breath-taking floral vignettes," as well as make "gains in social and educational importance." These predictions came to fruition. Most importantly for the Junior Gallery, the proceeds from the festival totaled $8,180, considerably more than the initial grant of $5,000. These funds would be used to buy more nature and art books, audiovisual equipment, and materials for the museum's art classes.[32]

The Collector's Corner, an innovation for the 1961 festival, added "eye-catching interest," according to reports. It was by far the most profitable section of the WGC show, netting $2,961.50, as compared to the $198 the WGC received from the sale of objects by local artists and $927 from the sale of 934 consigned items. Helen Beals, who chaired the 1961 Collector's Corner, may have initiated the concept, but whoever thought of it, it proved to be a brilliant idea. WGC members and friends donated all the items sold at the Collector's Corner, thus yielding 100 percent profit for the club. Not only were all the treasures donated, but they were donated at a

cocktail party held for WGC members, spouses, and friends. At the party, each member presented to the others the item she intended to donate to the Collector's Corner. Certainly no garden club member wanted to embarrass herself by donating something undesirable, damaged, or otherwise unsaleable. A fine art and antique dealer, Mr. Donaldson, volunteered his time to price the donations. The popular and profitable Collector's Corner returned as a feature of the flower festivals for years to come.[33]

For their next flower show, held on May 3, 1963, club members landed on the theme "The Chinoiserie Influence in English and European Gardens in the Late Eighteenth Century." An impressive chinoiserie-style gazebo, conceptualized by chairwoman Helen Sutherland, dominated Kirkwood Hall for the show. Helen's husband, Herman, sketched the plans for the structure and had Sutherland Lumber Company construct it. Plants, trees, and shrubbery, supplied by Evert Asjes Jr. of Rosehill Gardens landscape design and nursery, surrounded the gazebo. Some of the greenery was sold after the exhibition, as was the gazebo, which went to the historic Starlight Theatre. Once again, WGC members assembled flower arrangements to display in the loan gallery, furnished this time with eighteenth-century pieces and Wedgewood pottery from the Burnap Collection. Club members sold topiary trees and hanging baskets full of flowers they had grown, as well as terrace and garden furniture, decorative objects, and the items in the Collector's Corner. More than three thousand people attended the festival, a roaring success. The Westport Garden Club donated $11,022.31 to the Nelson-Atkins for the further beautification of Rozzelle Court. Thus, during a nine-year period, from 1955 to 1963, the WGC flower festivals provided the Nelson-Atkins with funds totaling $25,942.63.[34]

The Westport Garden Club was to hold two more extravagant flower festivals at the Nelson-Atkins. Jean Holmes McDonald chaired the seventh festival in 1966. As her assistant chair she chose Julia Tinsman, whom she described as a WGC member "with experience, taste, tact, talent, time, and patience." At the onset of the festival Mrs. McDonald stated, "Our idea was to cut down on the prodigious amount of work which had gone into it in the past—our fifty members were exhausting themselves." However, women of the WGC still managed to exhaust themselves putting on the 1966 show, as every member of the club served on at least one of the twenty-one committees and contributed to one of the fifteen different floral displays. The festival, with the theme "Decorating with Flowers Through the Centuries," was held on April 29. Members once again set up their flower arrangements in the museum's loan gallery, surrounded this time by furniture and objets d'art chosen by museum personnel to depict artistic styles during various time periods and places throughout the world, from Ancient Egypt to Italian Renaissance to American contemporary. In Kirkwood Hall a huge hot-air balloon, suspended from the twenty-two-foot-high ceiling, alluded to *Around the World in Eighty Days*, in keeping with the theme of art through the centuries and around the world. The four corners of Kirkwood Hall housed the Collector's Corner; the Flower Mart, which featured garden figures, ivy trees, hanging baskets, and pots handmade by Mr. Homic of Ironton, Missouri; the Children's Corner, which sold items made by the children in the Junior Gallery and Creative Arts Center; and the Luncheon *en Papillote* area. Andre's catered fabulous boxed lunches after both the museum coffee shop and Joe Gilbert "had turned down the *opportunity* to furnish the luncheon," according to Jean McDonald. Director Sickman hailed the 1966 WGC Flower Festival as "the most beautiful Garden Festival ever." When the accounts were all settled, President Laura Kemper Carkener presented a check for $11,042.80 to Laurence Sickman and James Seidelman, with the money designated to improve the educational facilities of the Junior Gallery and to further the beautification of Rozzelle Court.[35]

Laura Kemper Toll Carkener: A Talent for Leadership
Laura Kemper Carkener had the distinction of being the only member of The Westport Garden Club to have served as president twice. In her typical self-deprecating way, she dismissed her dedication and qualifications, saying, "Oh, I think they were desperate. No one else would do it." Not likely. In almost every organization with which Laura Kemper was affiliated, she went straight to the top.

Laura Kemper's mother was the third president of the Kansas City, Missouri, Junior League and went on to become vice president of the Association of Junior Leagues International, earning her a listing in the first edition of *Who's Who of American Women* (1958). When Laura Kemper joined the Junior League, she served in every executive position before her fellow members elected her president. Making

it a family tradition, Laura Kemper Carkener's daughter Laura Lee Carkener Grace, also a Westport Garden Club member, became the third generation of her family to serve as the Kansas City, Missouri, Junior League president. Despite holding high-profile positions, Laura Kemper found time to act as Prairie School PTA president and to chair many parent committees at the Sunset Hill School. She was also Jewel Ball chair and a trustee of the Kansas City Art Institute.

Figure 3.24 She was so capable that Laura Kemper Toll Carkener served as president of the WGC twice.

What, you may ask, did she do to relax? She gardened, of course, and not just in Kansas City. She held a dual membership in the Founders Garden Club of Sarasota in Florida, fondly remembered there for her gracious manner and willingness to chair any committee in need of a leader. Despite the calls of "Encore! Encore!," Laura Kemper's many accolades never went to her head.

Laura Kemper Toll Carkener was the last surviving founding member of The Westport Garden Club. She died in 2009 at the age of ninety-three. ❧

The Eighth Flower Festival Goes Dutch

The eighth WGC Flower Festival, also held at the Nelson-Atkins, took place in May 1969 and was chaired by Katherine "Kitty" Hall Wagstaff. The Wagstaffs hosted the Collector's Corner cocktail party, where WGC members proudly contributed offerings that included a Steuben vase, a Meissen lamp, Spode dinner plates, Bohemian glass urns, mirrors, silver, paintings, antiques, and other prized collectibles. The eighth festival relied again on Herman Sutherland and the Sutherland Lumber Company, which constructed a large, baroque windmill with rotating blades, announcing the Dutch theme of the event. In the loan gallery WGC members provided flower arrangements in the Dutch manner that complemented the museum's collection of sixteenth- and seventeenth-century furniture, Dutch still life paintings, pewter, and Delftware. The Flower Mart, Collector's Corner, commission shop, and luncheon combined to create a record profit of $13,645.80. A small percentage of the proceeds were reserved for a newly created Westport Garden Club Civic Endowment Fund, while the remainder went to the museum.[36]

Figure 3.25 Sutherland Lumber Company erected a large windmill for the Dutch flower festival of 1969, aided no doubt by Helen Sutherland. (WGCA)

Flower Festivals Bow Out

After the 1969 festival, past president and founding member Katie Kessinger reported that many club members believed they had devoted too much energy to flower festivals and flower arrangements at the museum; they now had other ideas for civic projects the club might undertake. We "disliked even temporarily giving up our association with the Gallery," but, Katie Kessinger wrote, "we should have our interests and our beneficiaries further afield."[37]

A change in museum policies provided an additional motive to move away from the floral extravaganzas at the Nelson-Atkins. After the Dutch-themed festival, certain of the museum staff had insinuated that some floral arrangements had led to insect infestations in the wooden frames of prized Dutch still life paintings. Although club members thought this highly unlikely, the concern caused the museum to rethink the idea of allowing the placement of floral displays near valuable paintings and furniture in the collection. In fact, the museum decided to tighten up its protocol in other areas as well, no longer allowing smoking, drinking, or eating in the museum's galleries.

The years of close connection between the Nelson-Atkins and The Westport Garden Club would not end though. WGC members would continue to support the museum and, eventually, create the Westport Garden Room, a room at the museum filled with plants, sculptures, and a fountain. Also, in 1985, after a sixteen-year break, the WGC would return to the museum to host a final flower festival. As Barbara Shackelford Seidlitz stated in her letter to Museum Director Marc Wilson, "All of our members consider it a unique opportunity, unequalled by that of any other Garden Club to hold this event at the Nelson-Atkins Museum and to have the advice of the curators and the assistance of your staff. We deeply appreciate their support and the cooperation of the Trustees."[38]

A WGC–Museum Liaison Forms

Over the years, Westport Garden Club members supported the Nelson-Atkins Museum of Art by providing flower arrangements for openings and receptions in addition to the financial aid accrued from the flower festivals. Director Sickman recognized the club's "incredible devotion" in his annual report of 1969, giving special recognition to WGC members Barbara Rahm and Peggy Sue Neal, "who with enthusiasm and seemingly inexhaustible energy, directed ten receptions and openings attended by some 5500 invited guests during this year." WGC members also received special praise for their work in preparing for the Gala Open House in 1969, celebrating the museum's thirty-fifth year. For the event, the entire museum was opened to fourteen hundred guests, who roamed the galleries filled with music and with flowers arranged by The Westport Garden Club.[39]

Eventually the museum recognized that a semipermanent liaison between the WGC and the Nelson-Atkins would help with the management of receptions, openings, and galas. The talented, generous, dedicated Barbara Rahm had been the go-to person for museum events for many years. She would continue to fill that role in a more official, if unpaid, capacity for many more years. Elizabeth Wilson, wife of the fourth director of the museum, Marc Wilson, remembers Barbara Rahm as "ever present" at the museum, her high heels clicking across the marble floors and her jangling bracelets announcing her presence as she busily attended to the organizing of events and the placing of flower arrangements composed by WGC ladies. According to Marc Wilson, Barbara never received the recognition or praise she deserved for the role she played at the museum. He credits her for being "the most charming general ever to enter the Nelson's door." She knew how to treat her troops, calling every person on the custodial staff and curatorial staff by name and treating each and every one of them with respect; they returned the favor. Eventually, Barbara Rahm was "promoted" to yet another unpaid position, assistant curator of special events. Finally, stretched to her limits, she retired in 1980, and a future WGC member, Ruth "Boots" Mathews Leiter, replaced her as a full-time, paid coordinator of special events. Then, when Boots relinquished her position in 1999, WGC member DeeDee Adams succeeded her until the pandemic of 2020 ended the seeming necessity or desirability for such a position.[40]

Barbara Forrester Rahm: A Floral Legend

High heels, coiffed red hair, a slash of red lipstick, and a wintertime fur fling—all fashionable accoutrements that made Barbara Rahm almost as dramatic as the towering floral designs she created for the Nelson-Atkins Museum of Art. When Barbara was serving as the museum's special events coordinator under Directors Laurence Sickman and Marc Wilson in the 1960s and 1970s, they considered her volunteer position so vital that she had her own office and special storage place, the "Rahm-a-torium." She was renowned for her creativity in elaborately arranging flowers and decor for exhibit openings, Society of Fellows events, small donor dinners, and many grand occasions.

Figure 3.26 The dedicated and talented Barbara Forrester Rahm served as WGC–Nelson-Atkins liaison for many years.

Barbara's volunteer job did not preclude her active involvement in the civic and social life of Kansas City or her duties as mother of adored twin girls. One of her daughters clearly remembers her mother donning a uniform, hat, and cape to swoop out the door for her Red Cross duties. In addition, Barbara chaired the Jewel Ball and helped found the BOTARs (Belles of the American Royal), a women's leadership organization that supports the mission of the American Royal in promoting the agribusiness economy of the Greater Kansas City area.

Barbara was generous with her time and abilities when it came to The Westport Garden Club, always willing to teach floral design classes, create spectacular arrangements for events, and participate in whatever projects required her talents. She was the first WGC member to receive the prestigious GCA Club Medal of Merit, in 1973.

What is the legacy of this gifted and vivacious woman? More than one person has said that it was her ability to make all of her assistants feel indispensable to the project at hand, big or small, and to treat everyone with whom she interacted, museum directors and museum janitors alike, in the same gracious manner. ❧

As the seedling club celebrated thirteen years as a member of GCA, Westport Garden Club members perceived that they had established their club as an important organization in the cultural and horticultural life of Kansas City. All involved felt it was time to branch out from floral designs and flower festivals to embrace other kinds of civic projects. Many members wanted to devote more time and effort to supporting GCA in its goals of protecting the environment "through educational programs and action in the fields of conservation and civic improvement." Of course, in order to fund any new project, WGC members would still have to raise money through more flower festivals or by other means. The club also vowed to keep in mind its original purpose: to share the love of gardening.[41]

3.27 *Left:* Education and hard work combine to produce gardens that exhibit beautiful color.

CHAPTER 4

The Westport Garden Club and the Environmental Movement in the 1960s and 1970s

THE ENVIRONMENTAL MOVEMENT, WHICH HAD started in the 1950s, gained ground in the 1960s and 1970s. GCA clubs were primed to support environmentalism, given their past success in restoring historic gardens, beautifying highways, saving redwood forests, and protecting flora and fauna. The GCA Conservation Committee encouraged every GCA club in the country "to translate the principles of conservation into personal action and thus encourage others to recognize our responsibility for our natural resources."[1]

WGC members responded. They contributed to GCA's National Student Conservation Program, sponsored children to attend Camp Hope (Kansas City's nature camp), and offered beautification and conservation advice, often unsolicited, to both the Kansas City, Missouri, Parks and Recreation Department and the Mission Hills, Kansas, Beautification Committee. WGC members also distributed "The World Around You" conservation information packets that GCA members Anna Rockefeller and Edna Fisher Edgerton had assembled in 1952, taking two hundred copies with them on a field trip to the Lakeside Nature Center in Swope Park for the Parks and Recreation Department.[2]

The WGC's major concerns and projects during the 1960s and 1970s, to be discussed in this chapter—legislation regarding pesticide control and water quality, the litterbug and billboard campaigns, the fundraiser for the restoration of the Wornall House grounds, and the environmentally focused 1971 zone meeting—all verify the club's involvement in GCA's

Figure 4.1 *Left:* Helen Jones Lea's garden. (Kilroy)

national conservation efforts. These projects were important, but two tenacious WGC members, Norma Henry Sutherland and Betty Clapp Robinson, led the way to seriously launch the club into the environmental and conservation movement with an ecology symposium. Norma was WGC's Zone XI (formerly zone 8) horticultural representative. Gleaning inspiration from an October 1973 presentation given to the WGC by the GCA horticulture chair, Mrs. Stephen Eberly Thompson of Portland, Oregon, Norma Sutherland partnered with Mrs. Thompson to develop the idea for a symposium. However, it was Betty Robinson who made the arrangements for the 1974 symposium and who would become vitally involved in GCA's environmental efforts.

Although club members became interested in these conservation topics, scrapbooks of the club's activities during these years indicate that garden tours, flower arranging, and horticultural studies remained of paramount importance to most of the members. Certainly, WGC women wanted to protect and preserve our country's natural resources, but rather than become die-hard environmentalists, most preferred to adhere to what they considered the primary purpose of The Garden Club of America, which was "to stimulate the knowledge and love of gardening."

The Influence of Rachel Carson's *Silent Spring*

Nonetheless, it was hard to ignore the clamor that arose after the publication in 1962 of Rachel Carson's meticulously researched book *Silent Spring*. Although many Americans, including WGC members, had voiced concern

about pollution and the country's misuse of natural resources, *Silent Spring* caused widespread alarm. Carson, a former marine biologist with the US Fish and Wildlife Service, conducted research that revealed how agricultural pesticides such as the widely used DDT (dichloro-diphenyl-trichloroethane) entered the food chain. DDT accumulated in the fatty tissues of animals (including humans) and caused cancer, genetic disorders, and even the decimation and extinction of certain plant and animal species. When *Silent Spring* was published, the chemical industry tried to discredit Carson, but her credentials as a trained biologist and her research placed her beyond reproach. She had found that a single application of DDT, even if diluted by rainwater, had a long-lasting effect in killing both targeted and untargeted insects. It also harmed birds and other animals and had negative implications for the nation's food supply.[3]

"Over increasingly large areas of the United States," Carson wrote in 1962, "spring now comes unheralded by the return of the birds, and the early mornings are strangely silent where once they were filled with the beauty of bird song." The image of the "silent spring" Carson painted struck a chord with the American public, helping them come to realize that nature is vulnerable to human intervention and that chemical "advancements" are not always positive. While some Americans had always cared about protecting the nation's forests and wildlife, up until the 1960s conservation had never raised widespread public interest. However, evidence uncovered by Rachel Carson—namely, that DDT could contaminate the whole food chain and cause the death of entire species—was too frightening to ignore. Now Americans in large numbers clamored for regulation of the chemical industry to protect the environment as well as their own health and well-being. After President John F. Kennedy directed the President's Science Advisory Committee to investigate the issues Carson raised in *Silent Spring*, DDT came under government scrutiny and was eventually banned. Americans across the nation considered Rachel Carson a heroine. In 1963, GCA awarded Carson a Special Citation for the profound effect her publication of *Silent Spring* had had on public consciousness. None other than Elizabeth Abernathy Hull was such a fan of Rachel Carson that she bought all of her books and bound

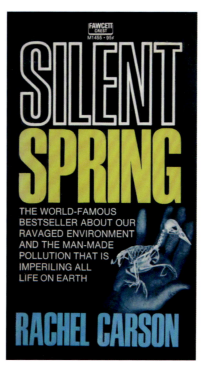

Figure 4.3 Rachel Carson's *Silent Spring* caused widespread alarm.

them in leather. (Bookbinding, as it turns out, was just one of Miss Hull's many skills.) After her death her "exquisitely bound" collection of Rachel Carson's books went to the GCA national library in New York City.[4]

The use of dangerous pesticides had long concerned the GCA Conservation Committee. In fact, Mrs. Rowland Robinson, the chair of the committee from 1959 to 1962, had listed pesticide control as one of the five important areas that required legislative action. Like their counterparts in other GCA member clubs, the women of The Westport Garden Club had warned the public about pesticide use even before the publication of *Silent Spring*. After the horrific floods of 1951, club members had taken a stand against supposed tree experts who were making profligate use of insecticides in swamped areas and on trees damaged during the storms. And while the pointed letters the club members sent to city officials had not eliminated the use of insecticides in the Kansas City area, they had indeed helped lead to the passage of a law requiring the licensing of those practicing tree surgery and treatment. Likewise, club members proselytized against the indiscriminate usage of pesticides and insecticides and

Figure 4.2 *Left:* The restoration and protection of the Kansas and Missouri prairie took on great importance for WGC members. (Kevin Sink)

distributed information about the dangers. It did not take any convincing for the WGC to lend its support to the GCA Conservation Committee's position on two pieces of legislation that came before the eighty-eighth and eighty-ninth US Congresses. One was the Pesticide Control Bill of 1964, which put stricter regulations on the former Federal Insecticide, Fungicide, and Rodenticide Act. The new bill, by closing loopholes that had permitted chemical companies to market and sell pesticides before fully testing them, was expected to safeguard the health and lives of the American people. A second important bill backed by the GCA and the WGC, the Water Quality Act of 1965, required states to establish water quality standards and uphold those standards regarding interstate water.[5]

Taking Aim at Billboards and Litterbugs

Highway beautification, which comprised eradicating billboards, stopping litterers, planting wildflowers, and eliminating roadside dumps, was also part of the conservation movement. The litter problem definitely resonated with The Westport Garden Club. The Federal-Aid Highway Act of 1956, which resulted in the Eisenhower Interstate Highway System, led to increased traffic and what seemed like a national epidemic of litterbugs. Since Interstate 70 from Kansas City to St. Louis was the first of the interstate highways built, it had the first and worst of the litter problems. While many GCA clubs across the country worked with schools to promote the use of litter bags in family cars, the Garden Club of Michigan went a step further and successfully petitioned the auto industry itself to either build trash receptacles into their cars or place litter bags in every car sold. This effort proved to be only partly effective. In the spring of 1957, The Westport Garden Club, under the guidance of President H.E. Hockaday, launched the "Litterbug Campaign" across Greater Kansas City. Members contacted the mayor, the City Council, the Chamber of Commerce, the Boy Scouts, and the school board to join in promoting the anti-litter program.[6]

Related to the litter problem was the ongoing national concern about billboards. When the media announced passage of the Billboard Control Amendment to the Federal-Aid Highway Act of 1956, headlines in many states read, "Garden Club Gals Out-Lobby the Billboard Lobby." However, the Billboard Control Amendment, or so-called Bonus Act of

Figure 4.4 GCA members banded together to try to out-lobby the billboard lobbyists. This Ogden Nash parody of the Joyce Kilmer poem "Trees" was also made into a postcard and quoted at garden club meetings across the country.

1958, was difficult to enforce since it was a voluntary act that allowed states to receive federal money if they did not allow construction of billboards on new sections of interstate highway being built within their boundaries. Billboards would continue to visually detract from the scenery along interstates in Kansas and Missouri. The Westport Garden Club and individual members wired their state senators, with few results.[7]

In his State of the Union address in January 1965, President Lyndon B. Johnson expressed his interest in a beautification program and even called a conference, the Conference on Natural Beauty, in May 1965, which was attended by eight hundred delegates. With voluntary outdoor advertising control due to expire on June 30, 1965, and visual pollution of highways and byways a growing concern, new legislation was needed. First Lady Claudia "Lady Bird" Johnson took the lead in calling for the passage

Figure 4.5 *Right:* Always an organization of tree huggers, the WGC launched itself into the environmental and conservation movement, in part to save such species as the little bluestem (*Schizachyrium scoparium*). (Sink)

of the Highway Beautification Act. This act became her pet project and was nicknamed Lady Bird's Bill. Signed into law by President Johnson on October 22, 1965, nine years after the passage of the Federal-Aid Highway Act of 1956, the Highway Beautification Act regulated advertisements on interstates and set standards for controlling junkyards and dumps.

Figure 4.6 Lady Bird Johnson attended the signing of the Highway Beautification Act, a.k.a. "Lady Bird's Bill," in the East Room of the White House on October 22, 1965.

Forty years after the passage of "her" bill, Lady Bird Johnson was honored at the City of Fountains Awards dinner, held on Tuesday night, April 26, 2005, at the GCA Annual Meeting in Kansas City. Although Mrs. Johnson could not attend the ceremony, having suffered a stroke in 2002, those assembled recognized her for her environmental work and, specifically, for her substantial efforts in support of the passage of the Highway Beautification Act, the first major legislative campaign ever launched by a First Lady. Lady Bird Johnson's views clearly matched those of the GCA and the WGC: Answers to the problems of growing urbanization, she said in a 1968 talk,

> cannot be found in [piecemeal] reform. The job really requires thoughtful interrelation of the whole environment. Not only in buildings, but

parks; not only parks, but highways; not only highways, but open spaces and green belts. A beautification in my mind is far more than a matter of cosmetics. To me, it describes the whole effort to bring the natural world and the manmade world to harmony. To bring order, usefulness, delight to our whole environment. And that of course only begins with trees and flowers and landscaping.[8]

Meaningful as it was, the Highway Beautification Act did not solve all the issues its supporters had hoped it would. The federal government did not allocate any funds for monitoring the billboard section of the legislation, so enforcement of the law fell to state governments, many of which failed to execute this portion of the law. Five years later, the Federal-Aid Highway Act of 1970 passed. This act authorized substantial funding for its programs as well as creating the Highway Beautification Commission as an oversight panel consisting of eleven members. Beginning in 1971, the commission would hold national symposiums to listen to local issues and raise national awareness. On this commission sat four US senators, four US representatives, and three members personally appointed by then President Richard M. Nixon, including GCA member Marion Ruth Thompson Fuller Brown, who had a Kansas City connection.

Marion Thompson was born in Kansas City and had graduated from Sunset Hill School before attending Smith College, marrying, and relocating to Maine. She went on to become a member of GCA's Piscataqua Garden Club, of York Harbor, Maine. She served three terms in the Maine state legislature, where she was the principal sponsor of Maine's Clean Water and Clean Air Act and of legislation that banned billboards in the state. She served on President Nixon's Highway Beautification Commission from 1971 until 1975. In 1977, she founded the Coalition for Scenic Beauty, now known as Scenic America, one of GCA's partners and the only national organization solely dedicated to advocating for reduction of visual pollution and preservation of scenic views. In 1988, GCA announced the approval of the Marion Thompson Fuller Brown Conservation Award for Small Flower Shows (now GCA Flower Shows). Her children established the award in honor of their mother's seventieth birthday and in acknowledgment of her leadership in conservation. In 2002, The Westport Garden Club, in partnership with the Anita B. Gorman Discovery Center, a Missouri

Department of Conservation nature center, received the Marion Thompson Fuller Brown Conservation Award![9]

Eight years after the passage of the Beautification Act, and two years after the Federal-Aid Highway Act of 1970 went into effect, the billboard problem persisted. In February 1973, the GCA Conservation Committee announced "a national campaign to urge the Governors and Highway Commissioners of all 50 states to proceed at once in carrying out the highway beautification program for their states." WGC President Kitty Wagstaff asked members to take part in the national letter-writing campaign, which they willingly did. Unfortunately, the billboard problem still plagues Missourians today, as anyone who has traveled I-70 will attest.[10]

A Powerful Committee Takes Shape

The Garden Club of America had never shied away from trying to influence legislators, knowing full well that they could not hope to meet their conservation goals without the support of legislation. That said, letter-writing campaigns could only go so far, so in 1968 the executive board of the GCA decided to form a committee, separate from its Conservation Committee, to relieve the Conservation Committee from having to lobby for and monitor legislation. The GCA's National Affairs and Legislation Committee (NAL), composed of five (and later seven) women, restricted itself to concerns consistent with the parent organization—namely, "protection of endangered species of flora and fauna; preservation and improvement of areas of horticultural, historic, scenic or ecological value; prevention of air, water, and soil pollution; and elimination of billboards in the interests of beauty and safety."[11]

As a former chairwoman of the GCA Conservation Committee, GCA President Willie Waller, who initiated the idea of forming the NAL, had a thorough understanding of the overwhelming work the Conservation Committee had tried to accomplish. Following her stint as president of the GCA, Mrs. Waller became the first chair of the NAL in 1969. She set ambitious goals for the new committee—to educate the entire GCA membership and the public about environmental legislative issues and to maintain a presence in Washington, DC, and be able, if necessary, to testify before Congress. With Willie Waller at the helm, the NAL set out to make a difference in conservation and protection of the country's natural resources.[12]

At the May 1969 board meeting of The Westport Garden Club, President Helen Sutherland read a letter from NAL Chair Waller in which she asked the WGC to express support for six pieces of legislation and to send wires and letters to legislators immediately regarding these bills:

S. 7, Water Quality Improvement Act: Wire Senator Jennings Randolph, Chairman, Subcommittee on Air and Water Pollution, Senate Committee on Public Works, with a copy to Senator Edmund Muskie, who introduced the bill

S. 1280, Endangered Species Bill: Wire Senator Warren Magnuson, Senate Committee on Commerce

S. 1753, Bill Prohibiting Sale and Shipment of DDT for Use in the United States: Wire Senator Gaylord Nelson, who introduced the bill

H.R. 9868, Identical to Preceding Bill, Introduced in the House by Congressman Bertram Podell: Wire the Hon. W. R. Poage, Chairman, House Committee on Agriculture

S. 1799, National Pesticide Commission Act of 1969: Wire Senator Allen J. Ellender, Chairman, Senate Committee on Agriculture and Forestry

H.R. 10504, Identical to Preceding Bill, Introduced in the House by Congressman Bertram Podell: Wire the Hon. W. R. Poage, Chairman, House Committee on Agriculture

Rather ambitiously, WGC charter member Maxine Goodwin moved that The Westport Garden Club wire its support for each of the bills described in Mrs. Waller's letter; the motion was seconded and carried. The WGC board also encouraged individual members to write or wire their congressmen and -women.[13]

Historic Preservation at the John Wornall House

As WGC members looked for a new civic project to replace their complete dedication to the Nelson-Atkins Museum of Art, Hazel Hillix Barton suggested that they help with the restoration of the 1858 John Wornall

House, located in Kansas City's Brookside area. More than a century old, the antebellum house had fallen into disrepair, and the Jackson County Historical Society was in the process of restoring it. The grounds of the house also needed rehabilitation. Seeing this project as an ideal combination of preserving an historic house and gardens and beautifying the city, WGC members decided to submit an application for a Founders Fund grant from GCA to support their efforts to raise money for the landscaping at the Wornall House. Established in 1934 in memory of Elizabeth Price Martin, the first president of GCA, the Founders Fund provides an annual monetary award ($5,000 at the time) to a project chosen by member clubs. The Garden Club of America *Bulletin* printed the WGC application for the Founders Fund grant in its January 1971 issue: "Landscape Restoration at the Historic Wornall House," proposed by The Westport Garden Club and seconded by the Garden Club of Denver. The copy read,

> In Kansas City, Missouri, the John Wornall House now being restored by the Jackson County Historical Society was one of the finest buildings in the area of Westport, the last important community on the pre–Civil War frontier at the head of the Oregon and Santa Fe Trails. Built in 1858 in the Greek Revival style brought into Missouri by plantation settlers from Kentucky, it is almost the only remaining brick farmhouse in the city which can be authenticated.[14]

Unfortunately, the project did not receive the Founders Fund award but was a runner-up. However, since WGC members had committed themselves to raising money to ensure the landscape restoration, regardless of whether they received a GCA grant, they went ahead with their project as soon as they had finished hosting their second zone meeting, scheduled for April of 1971.

The meeting concluded, WGC members resumed their focus on the Wornall House, under the direction of Hazel Barton, chair of the Civic Improvement Committee. The committee planned as a fundraiser a large garden party on the grounds of the historic house, set for spring 1973. Although the hope was to make this event far simpler than the elaborate flower festivals held at the museum, as it turned out, WGC members did not know how to do an event halfway. Indeed, every club member ended

Figure 4.7 Native grasses grow up around the John Wornall House Museum sign. (Kilroy)

up participating in what turned out to be a not-at-all-simple fundraiser. Inspired by a newspaper clipping describing the wedding of Roma Johnson and John Wornall, which had taken place in the parlor in 1865, club members decorated every room of the house with era-appropriate flower arrangements, along with Victorian treasures lent by club members and friends thereof. (One of those friends, the author of this book, who would become a WGC member, loaned an ornate silver pitcher with an engraved *W*, which was displayed on the dining table as a treasured "Wornall" family heirloom.)

The invitations, tickets, and decor all reflected a color scheme of pink and green. Strawberry-colored picnic baskets bordered in green held lunch fare typical of the nineteenth century, while another table presented glasses of sherry. Plants and topiaries, all of which club members had grown, were offered for sale on the south porch of the house. Meanwhile, the home's carriage house, besides housing the Collector's Corner donations—a tradition carried over from the flower festivals—gave contributors the chance to "buy" a brick to be used in replacing a Wornall House garden walkway that had fallen into serious disrepair. With all these innovative money-making opportunities and what turned out to be a beautiful day for the garden party, The Westport Garden Club managed to raise $10,412.44. The Jackson County Historical Society used the funds toward restoring the Wornall House and its grounds.[15]

Figure 4.8 WGC members decided to make their next project helping to fund the landscape restoration of the historic Wornall House. (Kilroy)

Figure 4.9 Westport Garden Party invitation, April 27, 1973.

Regional Ecology
Takes Center Stage at the 1971 Zone Meeting

In April 1971 the GCA Zone XI meeting had taken precedence over the Wornall House event. Libby Adams, WGC president, and Lillian Diveley, zone meeting chair, welcomed eleven national officers as well as sixty-one delegates from nineteen garden clubs in Zone XI. The theme of the three-day event, "The Heart of America Heritage," focused on the abuse of the region's ecological heritage and what was being done to restore its abundance. Dr. John P. Baumgardt, an esteemed botanist, University of Missouri professor, and immediate past president of the National Men's Garden Clubs, gave his views on the plight of the disappearing prairies, while Richard Rotsch of the Missouri Conservation Commission offered encouragement by describing ways in which the commission worked to conserve native plants.[16]

The event at Linda Hall Library began with a business meeting conducted by Mrs. Nicholas R. DuPont, first vice president of GCA; a conservation report by the GCA Conservation Committee chair, Mrs. Greeley; and a tour of the library and grounds. An afternoon at the Nelson-Atkins Museum of Art allowed delegates to see the plantings the club had done in Rozzelle Court, the GCA award-winning Junior Gallery, and the museum's renowned Chinese collection. Typical of all zone meetings, the WGC also hosted a flower show, which featured delegates' floral arrangements and horticultural specimens. Following the judging of the show, the delegates dined at The River Club, a private club known for its cuisine and for the view from its dining room of six counties and the confluence of the Kansas and Missouri Rivers.

The second day, Wednesday, April 21, 1971, the delegates attended a horticulture meeting at the Loose Park Garden Center and then had a chance to see the ongoing restoration at the Wornall House on their way to lunch. They spent the afternoon touring the gardens of six WGC members: Julia Bartlett, Lillian Diveley, Diana James, Margot Peet, Helen Thompson, and Julia Tinsman. That evening, cocktails were served at the estate-like home of Sally Ann Kemper and Thomas Wood. Following a zone Presidents' Council meeting on Thursday morning and a farewell luncheon at the home of WGC founding member Dorothy Cunningham, the meeting concluded. As with the first zone meeting, held in Kansas City in 1957, the delegates to the 1971 affair were awed by the city and the achievements

of the still-young GCA-member club. According to Meeting Chair Lillian Diveley, letters from delegates were "lyrically enthusiastic about the hospitality and success of the meeting."[17]

Figure 4.10 Zone delegates attended a horticulture meeting at the Loose Park Garden Center.

1974: Symposium on the
Ecology of Midwest Flyways and Prairie Flora

The next major project that The Westport Garden Club undertook—an ecology symposium—was a radical departure from flower shows and garden parties. Ultimately sponsored by The Garden Club of America and its Conservation and Horticulture Committees, the idea originated with Norma Sutherland. However, Betty Robinson, chair of the symposium, and Sally Cowherd, the WGC Conservation Committee chair, made the arrangements. The Westport Garden Club hosted the Symposium on the Ecology of Midwest Flyways and Prairie Flora on March 5, 1974, at the

Figure 4.11 *Left:* The 1971 zone meeting focused on the plight of the disappearing prairie. (Sink)

Figure 4.12 Delegates to the zone meeting were "lyrically enthusiastic about the hospitality and success of the meeting," according to Lillian Diveley. (WGCA)

Alameda Plaza Hotel. Open to the public, the symposium was intended to acquaint all attendees with the ecological problems in the Midwest and the need for legislation on various environmental issues.[18]

The WGC urged all interested conservationists and horticulturists in the region to attend the symposium. Also, many delegates came from the national organization, including Mrs. Nicholas R. DuPont, from Greenville, Delaware, who returned to Kansas City as president of The Garden Club of America; Mrs. Mart W. Reaves, GCA Conservation Committee chair, Dallas, Texas; Mrs. Leonard Kirby, Conservation Committee vice chair, Pebble Beach, California; Mrs. Stephen Eberly Thompson, GCA Horticulture Committee chair, Portland, Oregon; Mrs. W. L. Lyons Brown, GCA National Affairs and Legislative Committee chair, Harrod's Creek, Kentucky; and Mrs. Thomas M. Waller, past GCA president, Bedford Hills, New York. Other GCA delegates included Conservation Committee chairs and Horticulture Committee chairs from Zones IX, X, and XI.

The morning session of the symposium began with a focus on conservation, with the subject "Sanctuaries and Refuges of the Midwest Wetlands from Hudson Bay to the Gulf." The speakers included Charles Callison, executive vice president of the National Audubon Society, and Raymond C. Hubley Jr., executive director of the Izaac Walton League, one of the country's oldest and most successful conservation organizations. Following these presentations on wetlands, endangered species, and the history of the prairie, Ray Heady, retired outdoor editor of *The Kansas City Star*, spoke to attendees at the luncheon. The afternoon of the symposium was devoted to horticulture. WGC President Jean McDonald introduced Dr. John Baumgardt, the botanist and professor who had also spoken at the recent zone meeting. For the event, Baumgardt narrated his self-made film, *Wild Flowers of the Prairie*. Dr. Walter Kollmorgen, professor of geography at the University of Kansas, followed with a talk titled "The Uniqueness of the Prairie Environment and Diversity of Environment on a Microscale."

Exhibits in the ballroom at the Alameda Plaza included 1) a display presented by the Kansas Grassroots Organization, which presented working ranchers' suggestions for preserving the prairies; 2) slides of the history of the prairie from the Walt Disney movie *Vanishing Prairie*, shown by Tom Milne and Lucia Landon of the organization Save the Tall Grass Prairie (who also offered free literature and several kinds of prairie grasses and seed packets for sale); and 3) an exhibit put together by Shawnee Mission Northwest High School students, working with Wendell Mohling in cooperation with the federally funded program Project C.L.E.A.N. (Cooperative Learning through Environmental Activities in Nature). WGC members donated two hundred dollars to support the high school students' efforts to establish a ten-acre plot adjacent to their school, where they planted native grasses and plants. The students also developed a nature trail with interpretive signs describing man's interrelationship with the environment. The Westport Garden Club selected these projects, together called the Shawnee Mission Northwest Environmental Laboratory of Living, as the recipient of its 1974 civic award, hoping that the school's example would inspire other clubs, foundations, and interested individuals.[19]

Figure 4.13 Part of the morning session of the Symposium on the Ecology of Midwest Flyways and Prairie Flora dealt with protection of the Midwest wetlands, such as this area along Squaw Creek in northwest Missouri. (Lyndon Gustin Chamberlain)

Figure 4.14 Prairie flowers play an important role as pollinator plants, Dr. John P. Baumgardt explained as he showed his self-made film, *Wild Flowers of the Prairie*.

The great success of the symposium was largely attributable to Betty Robinson, who not only orchestrated the event but became vitally involved in GCA's environmental efforts, going on to serve as chair of the GCA Conservation Committee from 1977 to 1979.

Betty Clapp Robinson: A Voice for Conservation

A forthright woman with a "can-do" attitude, Betty Robinson always sat in the front row at membership meetings and "strongly encouraged" others to do likewise. Betty's interests and talents easily dovetailed with the mission and purpose of The Garden Club of America, especially in the fields of conservation and historic preservation. At a time when her club contemporaries were focusing on flower festivals and social events, Betty was quickly ascending the GCA leadership ladder, culminating in service as chair of the GCA Conservation, Nominating, and Admissions Committees. Ultimately, she was elected to the GCA Executive Board.

Figure 4.15 Betty Clapp Robinson was at the vanguard of the environmental movement.

In the vanguard of the environmental movement, Betty was the first WGC member to make these issues a priority, giving well-informed reports on current conservation topics at each meeting. She was instrumental in organizing the 1974 Symposium on the Ecology of Midwest Flyways and Prairie Flora, which was held in Kansas City—in conjunction with the GCA Horticulture and Conservation Committees. Her

strong commitment to environmental issues resonated with younger members and influenced the club's focus in succeeding years.

Betty also staunchly supported historic preservation, serving on the board of regents at Kenmore, the eighteenth-century home of George Washington's sister in Fredericksburg, Virginia. Toward the end of her GCA career, Betty served on the newly formed Archives of American Gardens (AAG) Committee. The AAG created a partnership with the Horticultural Services Division of the Smithsonian Institution to identify old glass lantern slides used by the GCA in the 1920s to identify members' gardens and garden images from the 1890s. The AAG Committee later became the GCA's Garden History and Design Committee.

Not surprisingly, the GCA recognized Betty with several awards, including the GCA Zone XI Conservation Award, the GCA Club Conservation Award, and the GCA Club Medal of Merit. ℘

A New Breed of Garden Club Women

The Symposium on the Ecology of Midwest Flyways and Prairie Flora encouraged WGC members' interest in environmental education, preservation of the prairies, and protection of the flyways. The symposium also inspired WGC members to take botanical field trips to remnant prairies. Interestingly enough, just as The Westport Garden Club embraced the environmental direction that The Garden Club of America was taking and the interest in birds that *Silent Spring* had rejuvenated, the GCA changed its mission statement, limiting "to share the advantages of association through conference and correspondence in this country and abroad" to this country alone and intentionally leaving out "birds." Many WGC members considered the latter deletion a mistake, as the symposium delegates had just heard from Charles Callison of the National Audubon Society about the importance of taking conservation measures to protect birds and other wildlife. However, GCA did add a very strong statement about restoring, improving, and protecting the environment through *actions*, not just words, and The Westport Garden Club did indeed spring into action.

The ecology symposium heralded a new direction for the club. A newspaper column published a decade beforehand had identified garden club members as "little old ladies in tennis shoes," according to *Mission Hills Squire* publisher Tom Leathers, but he took a different tack: "If you are like many, you probably think of a 'garden club' in terms of housewives, heliotrope, and hybrid roses." However, Leathers surmised, "this is a new era—and the garden club of today often comprises a new breed of woman, aware of our environment and concerned about our ecology." He went on to say that members of The Westport Garden Club "are trying to spread their awareness and concern to others. Their goal is to acquaint people from all walks of life with the need for legislation on different phases of the national environmental problems." The article also informed readers that the WGC symposium was just one of many across the nation sponsored by GCA during the past two years in an effort to get more people involved in environmental concerns. In conclusion, Leathers affirmed that "the 'garden clubwoman' of today is interested in much more than planting and pruning. She is doing her best to preserve this world of ours."[20]

Figure 4.16 The new breed of garden club women showed awareness and concern for ecology and maintaining areas like Dunn Ranch Prairie in northwest Missouri. (WGCA)

The WGC's Twenty-fifth Anniversary Marks the End of an Era

The United Nations designated 1975 as International Women's Year and organized a world conference in Mexico City on the status of women. The conference set goals to advance women's rights around the world and declared 1976–1985 as the UN Decade for Women. However, 1975 also marked the twenty-fifth anniversary of The Westport Garden Club.

On November 3, 1975, The Westport Garden Club held its anniversary celebration. Among the forty-nine members present were eighteen founding members, who personally remembered the challenges the club had met and its range of activities over the many years. Some of the changes were subtle. Although the mid-1970s was the age of bell-bottoms, tie dyes, and hippies, most of the WGC ladies still wore heels and suits with skirts to meetings. WGC members were still addressing each other as Mrs. or Miss So-and-So as well. Gone were the pillbox hats and white gloves, but pantsuits had become acceptable.

Newell "Honey Boy" Thornton and her daughter Prudence Thompson planned the anniversary event's program, which began with a report by Millie Hoover, the WGC's first acting president. Millie's presentation dovetailed into the presentation on the rest of the club's history, given by Lillian Diveley, who had prepared for her talk by reading all twenty-five years' worth of WGC minutes. Then Katie Kessinger, the first elected club president, gave a synopsis of the lavish flower festivals held at the Nelson-Atkins Museum of Art and the garden party at the Wornall House. All these events provided funds to support local beautification, horticultural, and conservation projects.[21]

Vivian Foster, who spoke next, said it was her privilege to report "on the civic projects to which Westport Garden Club has contributed time, talent, know how, and cash." The list was quite impressive: The Westport Garden Club had helped beautify an area on Quality Hill after the 1951 flood; planted flowering crabs and dogwoods at St. Luke's Hospital; landscaped the Westport Triangle; completed foundation plantings and added shade trees around the Conservatory of Music; and planted gardens at the Rehabilitation Institute, the Psychiatric Receiving Center, and the Kansas City Museum of History and Science. In Loose Park, the club had installed metal markers for tree species, contributed books to the Garden Center, and given trees to the Stanley R. McLane Arboretum. At Linda Hall Library,

WGC members had planted gardens and purchased and installed statuary and gates. In addition, club members had distributed GCA environmental packets to the Girl Scouts and the Lakeside Nature Center at Swope Park as well as contributed to the Laboratory of Living project at Shawnee Mission Northwest High School. Individual club members had served on the Kansas City Parks and Recreation advisory board and the Mission Hills Beautification Committee, volunteering their time and expertise.[22]

However, perhaps the greatest impact made by the women of The Westport Garden Club during its first twenty-five years was that made on the Nelson-Atkins Museum of Art, with which it had established a partnership in its early days. An itemized list of gifts that the museum had received from the garden club, as of November 1, 1975, totaled an impressive $53,332.41. Many of these gifts were the proceeds from the flower festivals held at the museum, all of which went toward education, the children's library, and the greening of Rozzelle Court. The highlight, of course, had been the WGC's role in creating the new Junior Gallery and Creative Arts Center, where children learned to combine art and nature, horticulture, and conservation—a project that won the GCA's esteemed Amy Angell Collier Montague Medal for the club's important contribution to the cultural growth of the community.

Last to speak at the twenty-fifth anniversary celebration was Jo Stubbs, then serving as the program chair for Zone XI. She spoke about "how well this club has carried out many of GCA's objectives." WGC members had been delegates to national and zone meetings, served as zone representatives, and given monetary gifts to support GCA programs such as the National Student Conservation Program. Several WGC members had won national and zone awards and citations in horticulture, garden design, and conservation in addition to awards at flower shows held at various venues.[23]

Josephine Reid Stubbs: An International Connection

A founding member and past president of The Westport Garden Club, Josephine "Jo" Stubbs remained a loyal club member until her death in 2006 at the age of ninety-eight. She wrote the first informal history of the club, which she presented at each new-member orientation. Her "grande dame" presence, coupled with her kindly disposition, prompted many a member to address her respectfully as "Mrs. Stubbs," even when formal titles were no longer commonplace in the 1980s and 1990s.

Jo was a fixture in the civic and social life of Kansas City, serving as president of the Children's Relief Association at Children's Mercy Hospital, of the Visiting Nurse Association, and of The Barstow School board. She was Jewel Ball chair in 1957.

Figure 4.17 A founding member and past president of the WGC, Jo Reid Stubbs took great interest in the Rome Prize, a GCA scholarship offered at the American Academy in Rome.

Jo was asked to serve at the zone level in several GCA positions, and, like many garden club members, she was drawn to one of the GCA's specific areas of interest. In her case, it was scholarship. The GCA offers over $300,000 in scholarships each year in fields related to its mission. The oldest and most prestigious of these is the Rome Prize at the American Academy in Rome (AAR). Fully funded by GCA, the Rome Prize offers one individual the opportunity to spend a year in Rome at the Academy's Villa Aurelia headquarters to pursue studies in landscape architecture. There, landscape architecture students are joined by twenty-eight other scholars in the fields of writing, music, and art. Jo served as GCA's liaison to the AAR, sat on the selection committee, and served as a trustee and trustee emerita, a singular honor. Jo never sought recognition for all her devoted work, nor, sadly, did she ever formally receive it. ✂

For its part, the "men's auxiliary" of the WGC had pitched in to support and add to both the creative and social aspects of the club. Husbands and significant others were often included on garden tours and social events, and all of them were lauded at this anniversary for having suffered through long telephone conversations and the concept of flowers having priority over preparing dinner. However, Lillian Diveley gave special recognition to 1) Frank Bartlett, for his interest and generosity in allowing The Westport Garden Club to hold its meetings at the Linda Hall Library; 2) Larry Sickman, for his assistance and guidance of the many WGC activities held at the Nelson-Atkins; 3) Phil Rahm, for his horticultural contributions, including the use of rock mulch on tomato plants; 4) George Gordon, for his assistance drawing up the WGC incorporation papers and facilitating the club's achievement of tax-exempt status; 5) Web Townley, for his guided "preservation tour" field trip to the River Quay; 6) Jack O'Hara, for his design of the packets for the Symposium on the Ecology of the Midwest Flyways and Prairie Flora; 7) Fred James, for his designs for the dogwood needlepoint badge and for festival flyers; and 8) Herman Sutherland, for his "good nature and ingenious creativity in the themes and visual props of our elaborate festival decors."[24]

Many WGC members viewed the twenty-fifth anniversary celebration as the end of an era of having committed themselves almost completely to flower festivals, flower arranging, and gardening. It was time to embrace other aspects of the larger GCA, and the club was poised to do just that. WGC members began to emerge as examples of the "new breed" of garden club women, not only as gardeners and flower arrangers but as horticulturists, environmentalists, conservationists—women who could make a difference in the wider world around them. ❧

Figure 4.18 *Right:* Helen Lea's garden. (Kilroy)

CHAPTER 5

Educational Opportunities Expand

THE LOGO OF ANY ORGANIZATION is meant to symbolize something particularly important to it. In 1921, eight years after the formation of GCA, Henrietta Maria Stout, one of the founders of the Short Hills Garden Club in New Jersey, designed what was then called the GCA "insignia." It focused on the two guiding principles of the organization: education and horticulture. In the center of the logo, the lamp of enlightenment symbolizes education, while the surrounding oak leaves and twelve acorns represent horticulture and the twelve original clubs, which included Short Hills Garden Club. The present-day logo, which was modernized by artist Marilyn Worseldine in 2002, has the same symbolism but is digitally adaptable.

Education remains essential to the work done by all GCA clubs. In the club's statement of purpose, which the GCA rewrote in the 1970s, education is mentioned twice, with an emphasis on "*educational* meetings, conferences, correspondence and publications" as well as restoring, improving, and protecting "the quality of the environment through *educational* programs." In the 1980s, the GCA increasingly began to see its role as an environmental educator, of both its own members and the public. Environmental concerns like water and air pollution and endangered plants and animals cried out. Now it was time to address the underlying causes of these environmental crises. Reaching out to its member clubs across the country, GCA encouraged them to take action and make their voices heard.[1]

Figure 5.1 *Left:* The creeping ivy (Hedera helix) flowers on the shagbark hickory tree in Sharon Wood Orr's backyard attract monarch butterflies. (Sharon Orr)

Figure 5.2 A lamp of enlightenment on the GCA logo symbolizes education, while the oak leaves represent horticulture.

THE GARDEN CLUB *of* AMERICA

When Norma Sutherland took office as WGC president in 1977, she sought to "bring new life and interest and a closer adherence to the goals of The Garden Club of America." At the time, though, a considerable contingent of WGC members already held GCA leadership positions. Julia Tinsman was the Zone XI Awards Committee representative, and three other WGC members held important GCA offices: Diana James was vice chair of the Horticulture Committee, Betty Robinson was chair of the Conservation Committee, and Libby Adams had held three GCA positions consecutively and would become a director the following year. During Norma's presidency The Westport Garden Club moved more in line with GCA objectives. Without the consuming job of planning and executing flower festivals—eight in fourteen years—WGC members could take action on new projects and educate themselves and the public about horticultural and conservation issues.[2]

In 1981, when GCA announced its national awareness campaign to save endangered American wildflowers, the Woodside-Atherton Garden Club of San Francisco launched the program by sending postcards to residents of that area to make them aware of an endangered plant native only to the area surrounding San Francisco Bay—the fragrant fritillary lily (*Fritillaria liliacea*). The club's campaign to save the fritillary lily aided the cause; today the lily, though still threatened, is off the endangered list. Of the 184 GCA clubs, 147 followed the San Francisco club's lead in bringing attention to wildflowers, including the WGC, which chose to increase public awareness of the endangered large-flowered, or so-called showy, beardtongue (*Penstemon grandiflorus*). This native penstemon, which was known to grow only in grassy, open, loess-soil areas in the northwest corner of Missouri, is rare and therefore not as well known as its cousin *Penstemon digitalis*, or foxglove beardtongue. Hoping to generate interest in cultivating and saving this drought-loving prairie species, WGC members sold postcards featuring showy beardtongue. The campaign proved successful; today many nurseries sell *Penstemon grandiflorus* seeds, and the plant, which is still considered endangered in some states, is no longer on the endangered list in Missouri.[3]

In 1989, GCA's National Affairs and Legislation (NAL) and Conservation Committees introduced position papers to identify the legislation, policies, and individual action that the national organization supports. These issues include clean air, clean water, climate change, native plants, waste management, sustainable agriculture, and land conservation. When Suzanne "Susie" Slaughter Vawter became WGC president in 1989, she made it the responsibility of the club's Conservation Committee chair to keep club members informed about position papers and the status of pertinent legislation so that the WGC could join the national organization in advocating for a beautiful and healthy planet.

While the WGC moved closer to adhering to the goals and priorities of the GCA during the 1970s and 1980s, it was not until 2013 that the club wrote a mission statement that reflected its role in horticulture and conservation action and education. At the annual meeting in 2013, the

GCA had urged each member club across the nation to develop its own strategic plan to internalize the national organization's vision, goals, and mission. The WGC took immediate action and chose Cindy Cowherd to head its Strategic Planning Committee. All readily agreed on their club's new vision: "to enhance, protect, and preserve the natural beauty of our community." The committee set five goals: 1) to implement an all-club fundraiser at least every other year, 2) to hold an educational or flower event at least every other year, 3) to provide optional workshops for club members in the fields of horticulture, floral design, and photography, 4) to have at least one membership meeting per year that included a hands-on workshop, and 5) to implement a plan to encourage camaraderie and involvement.[4]

To compose a new mission statement for the WGC, Cindy said the committee used the GCA purpose statement as its model but simplified the language to reflect the Kansas City club's special emphasis on the importance of education. At the September 2013 WGC meeting, President Nancy Lee Smith Kemper read the new statement: "The mission of The Westport Garden Club is to share the love of gardening through *education*, action, and creative expression in the fields of horticulture and conservation." The members readily approved it.[5]

With education of paramount importance to the WGC, each club member participates in one or more of the club's standing and special committees, in so doing learning about the various roles played by both the WGC and the GCA in such areas as horticulture, conservation, garden history and design, and photography. One committee especially committed to education is the Scholarship Committee, the members of which seek qualified applicants who can take advantage of the twenty-nine merit-based scholarships and fellowships that GCA offers. In 2018, Scholarship Committee Chair Kathy Garrett Gates announced at a club meeting that for two years in a row, the committee had succeeded in recruiting worthy candidates for GCA scholarships, including the recipient of the first ever Freeman Scholarship for Native Plant Study.[6]

Figure 5.3 *Right:* With a postcard campaign, the WGC increased public awareness of the endangered large-flowered beardtongue (*Penstemon grandiflorus*).

Apart from committee work, the club's monthly programs, creative workshops, conferences, and presentations by visiting specialists offer WGC members many opportunities to learn, while field trips enable members to become better acquainted with the land around them.

Programs, Workshops, Conferences, and Publications

Under its new vision of expanding education, monthly programs provide WGC club members (and sometimes invited guests) the opportunity to learn from expert outside speakers. For example, in 2015 Kevin Sink spoke on nature and garden photography. Wendy Paulson came from Chicago in 2016 to present a program about volunteers who are restoring biodiversity to Chicagoland's "wilderness," and Tom Schroeder presented "Assisting Native Bees in the Garden." Also in 2016, Linda Hezel from Prairie Birthday Farm, a small family farm in Clay County, Missouri, shared her knowledge about growing sustainably produced fruit, vegetables, herbs, and edible flowers, as well as producing eggs and honey. Mary Nemecek from Burroughs Audubon Society of Greater Kansas City presented a program in 2017 entitled "Support Conservation in Your Community."

In January 2018, the WGC invited John Gordon to speak about BoysGrow, a program for "mentoring Kansas City's urban youth through agricultural entrepreneurship." Gordon established BoysGrow in 2010, after having witnessed the life-changing experience of working on a farm for a fourteen-year-old boy in the juvenile justice system. The teenage boys who apply for this program commit to working in it for two years, three days a week during the summer and twice a month during the school year. The boys are bused from downtown Kansas City to a farm twenty-five minutes away, where they choose to work on one of three core teams: farming, culinary arts, or construction, and on a supplemental team that involves mechanics, landscaping, animal husbandry, marketing, and public speaking. The boys earn a paycheck for their work in the program and acquire social skills including teamwork, flexibility, integrity, and responsibility, which help them go on to higher education and/or to rewarding careers as chefs, farmers, or businessmen.

Gordon's presentation made such an impression on WGC members that they decided to make BoysGrow the beneficiary of their J. McLaughlin shopping days in May 2018, and they successfully proposed BoysGrow for

Figure 5.4 The BoysGrow program, established in 2010 by John Gordon, provides a life-changing experience for urban teens. (BoysGrow)

Figure 5.5 Some boys in the BoysGrow program choose to work in culinary arts—one of the three core team choices, along with farming and construction. (BoysGrow)

the GCA Club Civic Improvement Commendation, granted later that year. Since then, BoysGrow has been featured by Norah O'Donnell on the *CBS Evening News* and on the show of celebrity cook Rachael Ray. Television chef Lidia Bastianich, renowned for her New York restaurants Felidia (which closed in 2021) and Becco, as well as Lidia's in Kansas City, also has generously supported BoysGrow.[7]

Workshops, at which coffee and good conversation always flow, provide a wealth of educational opportunities for club members and cement friendships. A bulb workshop in November allows members to get together to plant amaryllis and paperwhites to bloom during the holidays. Occasionally the club holds a workshop on the propagation of trees, and several members now have trees in their yards—ranging from seedlings to mature—that they started at one of these workshops. In addition, study groups allow WGC members to pursue new interests and learn new tricks. Kathy Gates deserves credit for starting a Horticulture Study Group in 1994. Blair Hyde formed a monthly flower-arranging workshop in 2006 and even outfitted her walk-out basement to accommodate several work areas. The Flower Study Group offers multiple workshops, as does the Photography Study Group.

WGC members have several learning opportunities through attending GCA conferences in person, and since the pandemic, GCA members can also experience most conferences online. Each club sends two delegates to GCA annual meetings and zone meetings, and one lucky WGC member attends the annual horticulture conference named for Shirley Meneice. A member of the Carmel-by-the-Sea Garden Club in California, Meneice became known as one of GCA's most revered horticultural icons. In 2002, she organized a GCA horticultural meeting at the Huntington Botanical Gardens, which proved to be so successful that it became an annual GCA event held in various locations. When the GCA Executive Committee voted in 2003 to name the meeting for her, Shirley Meneice was surprised and flattered. She attended the conference named for her every year until her death in 2020 at age ninety-seven.

Another conference that one or more WGC delegates attends each year is the National Affairs and Legislation (NAL) Conference, held in Washington, DC. Delegates learn about updates on legislation and policies related to conservation. In February 2004, besides having learned "the scoop" on the current energy bill, delegates Dody Phinny Gates and Alison

Figure 5.6 The bulb workshop each fall provides an educational and social opportunity for club members, including Alison Wiedeman Ward, Kristie Wolferman, Janie Hedrick Grant, and Dody Phinny Gates Everist. (WGCA)

Bartlett Jager returned from the NAL conference with a humorous story. They had attended a presentation by then Senator Hillary Clinton, who began her speech with a gardening anecdote: "I love getting my hands in the dirt when planting my bulbs in the spring." This comment might have worked with another group, but not with members of GCA, who know that bulbs need to be planted in the fall![8]

Besides learning new information at GCA conferences, club members also have access to several GCA publications. The *Bulletin*, published since 1913, is currently issued quarterly and distributed to all GCA-affiliated club members. GCA also publishes the following triannually: *By Design*, a twenty-four-page, full-color magazine produced by the Floral Design

Figure 5.7 Shirley Meneice Horticulture Conference delegates Peggy Kline Rooney and Pam Sutherland Gyllenborg talked with Shirley Meneice at the last conference she attended before her passing in 2020. (Photo courtesy of Pam Gyllenborg)

Committee; *ConWatch*, an online magazine produced by the Conservation Committee; *Focus*, a newsletter written by the Photography Committee; and *The Real Dirt*, a newsletter of gardening tips and in-depth horticultural articles published by the Horticulture Committee. In contrast, the GCA publication *Call to Action* is released as needed, often with time-sensitive information urging GCA members to contact their legislators about certain bills or to otherwise take action.

In August 1994 the WGC began issuing its own publication, *La Belle Jardinière*. Originally published as a four-page magazine, *La Belle Jardinière* in 2018 began taking the form of an online newsletter. With beautiful photographs of events and field trips, notes of congratulation, and schedules of events, club members consider the publication a treasure trove of information and memorabilia. Interestingly enough, at least three editors in chief of *La Belle Jardinière* have gone on to the presidency of the club, including Margaret Weatherly Hall, Laura Babcock Sutherland, and Mary Ann Huddleston Powell.

Figure 5.8 The WGC began issuing its own publication, *La Belle Jardinière*, in August 1994.

Another useful source of information, with a calendar, meeting minutes, a member directory, and much more is the WGC website itself, which Nancy Newell Stark and Kristin Colt Goodwin set up and demonstrated to club members in November 2011.

Field Trips Far and Wide

When in 1974 The Westport Garden Club hosted the Symposium on the Ecology of Midwest Flyways and Prairie Flora, many club members did not know very much about the ecology of the region. The symposium whetted their appetites to learn more. Thus, the club began to plan field trips to view different aspects of the Midwest's topography so they could better understand the conservation issues.

The first of several informative field trips took place in fall 1976. The program chairs, Jo Stubbs and Marie Bell Watson O'Hara, arranged a visit to Maple Woods, located on the outskirts of Kansas City. The previous spring, real estate developers had threatened to bulldoze this unique 39.3-acre area of mature sugar maples and burr oaks. The Nature Conservancy stepped in to advocate for the preservation of Maple Woods, a successful effort to which the members of The Westport Garden Club donated $500. Upon visiting the site firsthand, club members took pride in seeing what their contribution had helped preserve. At the time of their visit, this old-growth timber forest was resplendent with fall color and a plethora of forest birds. A picnic, replete with wine, arranged by Marie Bell and Jo made the day even more special. Now recognized by the National Park Service as a significant "Natural Area," Maple Woods is maintained by the Missouri Department of Conservation in cooperation with the City of Gladstone and The Nature Conservancy.[9]

Figure 5.9 Having donated money to save Maple Woods from real estate development, WGC members toured the woods in fall 1976. (Wolferman)

In May 1977, WGC members and guests, including a visiting Zone XI Horticulture Committee representative, Mrs. James Dern of Winnetka, Illinois, took a bus trip from Kansas City to visit Missouri prairies and glades. On board the bus as their guide and lecturer was Dr. John P. Baumgardt, the esteemed botanist and University of Missouri professor who had participated in the ecology symposium in 1974. He explained that once, prairies had covered more than a quarter of the state of Missouri, but in 1977 less than one half of one percent of the original prairie remained. The first stop of the carefully planned botanical field trip was the McGinnis Farm, a privately owned property of remnant prairie located fifty-five miles south of Kansas City. The McGinnis family manages its prairie acreage by removing invasive species and allowing Black Angus cows to graze on their land, resulting in grass-fed beef.

Figure 5.10 In May 1977, garden club members and guests took the first of several bus trips to visit Midwestern remnant prairies. (WGCA)

Next on the agenda was the Taberville Prairie Conservation Area, located in St. Clair County and one of the largest remaining remnant prairies in Missouri. The Missouri Prairie Foundation, founded in 1968 to conserve Missouri's prairies and other native grasslands, uses prescription burning and controlled grazing to manage Taberville's 1,360 acres of tallgrass prairie. On this trip, WGC members viewed wildflowers in their prime and saw some of the birds and animals that are specially adapted for life on the open prairie, including greater prairie-chickens, northern bobwhites (quail), regal fritillaries (butterflies), and slender glass lizards. After a full day in the field, WGC members spent a night at the farm of club member Helen Sutherland and her husband, Herman, in Maysville, Arkansas, before heading the next day to Wolf Pen Gap, Arkansas, a park area featuring glades, high mountain vistas, beautiful waterfalls, and mixed pine and hardwood forests. Having experienced two botanical zones—prairie and Ozarks—the women and their guide found the bus ride home to be a bit of an experience, too, as they traveled under tornado alerts. All aboard deemed it a memorable field trip.[10]

Lone Prairie Park in Lawrence, Kansas, was a day-trip destination in October 1978, followed in 1979 by an overnight bus trip to Bartlesville, Oklahoma, to tour the Osage prairie. This trip, which included both WGC members and their husbands, was educational but also so much fun that the club planned another overnight field trip for the fall of 1980—to the Flint Hills. This area of tallgrass prairie, located in eastern Kansas and north-central Oklahoma, is also known as the Bluestem Pastures. The Flint Hills got their name from the brigadier general and explorer Zebulon Pike, who in 1806 wrote in his journal, "passed very ruf [sic] Flint Hills." The tallgrass prairies of the Flint Hills were once part of a massive, 150-million-acre tallgrass prairie that extended from present-day Texas to Canada. Unlike other tallgrass prairies, typically known for their deep soil base, the Flint Hills have a shallow soil base sitting on abundant residual layers of limestone and chert (or flint) deposited by an inland seaway millions of years ago. In fact, the surface layers of limestone and chert stopped early settlers from farming the land, and they turned instead to cattle ranching, still the biggest industry in the area.[11]

Figure 5.11 *Left:* Club members toured Taberville Prairie in western Missouri and viewed wildflowers in their prime. (Sink)

Bound for the Flint Hills, WGC members and husbands boarded a bus on the afternoon of Friday, October 28, 1980. After an overnight at the Ramada Inn in Manhattan, Kansas, the first stop was the Konza Prairie Biological Station, an experimental project founded by The Nature Conservancy. Wearing casual attire and good walking shoes, the trip participants walked the prairie area and listened to experts explain the importance of preserving tallgrass prairie land. Back on the bus, the next stop was the cattle ranch of WGC member Kitty Wagstaff and her husband, Bob, where the group had lunch and enjoyed "shopping" the area for dried grasses and plants to use for fall arrangements. The trip home traveled the Skyline Drive, a gravel road from Alma to Paxico, where the unique slopes of the Flint Hills and the beautiful views of creeks, birds, and wildflowers seemed like another world before the bus hooked back up with I-70 for the rest of the ride home.

In November 1981, WGC members and spouses again boarded a bus, this time to return to the Oklahoma prairie, but four years later, in October 1985, WGC members took a different kind of field trip. Carolyn Bond, the First Lady of Missouri, had invited club members to come to Jefferson City to view the newly restored governor's mansion and surrounding gardens. She also served them a "delicious white wine" and lunch.[12]

Also in 1985, WGC members planned a field trip to the newly expanded treatment facilities at Water District #1 in Johnson County, Kansas, as part of a three-year water study program initiated by GCA. The "dry run" of the water facility tour, according to Conservation Committee Chair Jeanne McCray Beals, was "quite fascinating." However, when the tour finally took place in March 1986, Jeanne said the pre-tour talk became a bit "windy . . . and there was no time to tour." Program Chair Betty Thompson, who had planned the trip and gone on the "dry run," heard about the disappointing field trip, but she "was thankfully in Jamaica." In 2024, with water conservation and pollution again a major source of concern for GCA members and for the general public, Conservation Committee Co-Chairs Laura Woods Hammond and Jennifer Ball Jones organized another educational tour to a water-treatment plant—Johnson County Wastewater, located in Leawood, Kansas.[13]

In September 1986 Doug Ladd, a biologist with The Nature Conservancy and an expert on prairie and forest plants, made a presentation to WGC members. Ladd talked about the importance of conservation, land management, and the ecological restoration of tallgrass prairies in order for native species of plants and animals to survive. His inspirational talk caused club members to plan a bus trip to the Grassland Heritage Foundation in Olathe, Kansas, to see firsthand an example of prairie preservation and to gain a better understanding of regional ecological concerns. They also discerned the need for funding organizations that could help preserve the remnant prairies and endangered plants. In her 1987 year-end report, WGC President Marie Bell O'Hara wrote, "Our most significant contribution of the year was our pledge of $2500 to the Missouri Botanical Garden for the Center for Plant Propagation to defray the costs of collecting endangered plant species. We are proud of this contribution which indicates our concern for the future." She also reported that the WGC had made donations to the Grassland Heritage Foundation, the Save the Redwoods League, the Student Conservation Association fund, The Nature Conservancy, and the Missouri Prairie Foundation.[14]

Figure 5.12 Marie Bell Watson O'Hara designed and maintained her own herbaceous perennial gardens. (WGCA)

Marie Bell Watson O'Hara: Gardener, Teacher, Golfer

"I just wanted to be outdoors!" exclaimed Marie Bell O'Hara when asked how she had come to love gardening. A Westport Garden Club member for over fifty years, Marie Bell designed and maintained her own extensive herbaceous perennial gardens. She generously shared divisions and cuttings with fellow gardeners, and many members still grow the progeny of her plants in their gardens today. Whenever GCA had its plant exchange, Marie Bell's greenhouse sheltered the newly propagated plants until their journey to the annual meeting.

Marie Bell also is well known for her large, prize-winning traditional flower arrangements. She often assisted Boots Mathews Leiter Churchill, the Nelson-Atkins Museum of Art coordinator of special events and a WGC member, with arrangements for museum events. Marie Bell was a sought-after teacher of floral design, always encouraging and enthusiastic, never judgmental.

"Just wanting to be outdoors" for Marie Bell included playing a highly competitive game of golf. She held the course record at The Kansas City Country Club from 1956 to 2016 and won the club championship eleven times. She was also a city champion.

Apparently, talent runs in the family: Her nephew is golf legend Tom Watson.

Marie Bell is also a classical pianist and for years performed for musical groups and other organizations. Her late husband, Jack, was a well-known Kansas City artist whose watercolors hang in several museums. Charming and affable, he often assisted the club when called on for his artistic skills.

Marie Bell served as club president and received both the GCA Club Floral Design and Horticulture awards, as well as the GCA Club Medal of Merit. She was succeeded in her term as club president by her sister-in-law Sallie Bet Watson, who was also a WGC star. ∽

In 1988, WGC members took the first of what turned into many field trips to visit the newly opened Powell Gardens, which launched a partnership based on both a shared purpose and many connections between WGC members and the Gardens. A field trip to the McGregor Herbarium at the University of Kansas in 1992 included lunch at the Boots Adams Alumni

Center, a lecture by Dr. Kelly Kindscher on the grasses and land configuration of Kansas, and a special program by Craig Freeman titled "Wild Flowers of the Southern Plains."

Figure 5.13 In 1988 WGC members took the first of many field trips to Powell Gardens, just east of Kansas City. (Powell Gardens Archives [PG Archives])

In 2001, WGC members Jean Holmes McDonald Deacy and her daughter Virginia "Gina" McDonald Miller invited the club to visit their Mashed O Ranch in the Flint Hills of Kansas near Cottonwood Falls. Four years later, in 2005, Gina and her mother provided the opportunity for delegates at the GCA Annual Meeting in Kansas City to take a trip to the ranch, see the unique ecosystem of the long-grass prairie, and witness a managed burn. In 2008, club members again boarded a bus to visit the Mashed O Ranch and the Tall Grass Prairie National Preserve, where they saw another controlled burn. Gina's dedication to conservation earned her a GCA Zone XI Conservation Award in 2002 and a GCA Club Conservation Award in 2007, as well as an award from The Nature Conservancy of Kansas in 2014.

On October 2, 2012, the WGC headed to Salina, Kansas, to visit The Land Institute. Founded as a nonprofit organization in 1976, The Land Institute works to rethink and redesign agricultural methods to make them compatible with soil and water conservation. According to Executive Director Dr. Wes Jackson, "Our work is dedicated to advancing perennial grain crops, which can help build and protect the soil. The Land Institute is committed to researching and developing food production methods that sustain the land and soil, a precious resource in an increasingly precarious state around the globe." The WGC had earlier proposed Dr. Jackson for a GCA national medal, and in 2012 he received the Elizabeth Craig Weaver Proctor Medal "for exemplary service and creative vision." The WGC feted Jackson at a June 6, 2012, dinner held at the home of Marilyn Bartlett Hebenstreit and her husband, Jim.[15]

In September 2014, WGC members took a bus trip to The Nature Conservancy's Dunn Ranch Prairie—more than three thousand acres of remnant prairie near Hatfield, Missouri. Dunn Ranch is in what The Nature Conservancy calls the Grand River Grasslands region, which encompasses eighty thousand acres located in southern Iowa and northern Missouri. This area, in turn, is part of the much larger tallgrass prairie region, which at one time stretched across the Midwestern states from Kansas to Ohio and from Canada to Texas. The Nature Conservancy hopes to keep the Grand River Grasslands, which currently lie in both the private and public domain, as a refuge for native plants and animals, a concept that will take a good deal of education and cooperation.[16]

In 2015, WGC club members visited the Prairie Garden Trust, an eight-hundred-acre plot of land south of Fulton, Missouri, owned by retired physician Henry Dombe. Photographer Kevin Sink gave club members a tour of the photographic vantage points on the property on May 30. After an overnight in Fulton, George Yatskievych, a botanist at the Missouri Botanical Garden and one of the leading authorities on Midwestern wildflowers, identified flora on a second Prairie Garden Trust tour.[17]

After nearly four decades of field trips to prairies, WGC members had learned a lot about conservation and the importance of maintaining prairie remnants. Therefore, when the members learned about Jerry Smith Park, the last remnant native prairie within the limits of Kansas City, they enthusiastically decided to make protecting it one of their missions. A 2016 botanical study of the park provided scientific proof of the biological integrity and quality of Jerry Smith Park's prairie. WGC members used this evidence to forge a collaboration with GCA's Partners for Plants, a program that provides guidance and monetary grants to projects in communities that are

Figure 5.14 Dunn Ranch Prairie includes more than three thousand acres of remnant prairie.

Figure 5.16 In 2014, WGC members, including Wendy Jarman Powell and Marilyn Bartlett Hebenstreit, explored Dunn Ranch Prairie. (WGCA)

in line with GCA's mission, purpose, and vision. Field trips to Jerry Smith Park became commonplace as groups of WGC members pitched in, helping conservationists collect native seeds and eradicate invasive plant species.

Other field trips focused on birds and bird migration. In 2014 the club made a journey to the National Audubon Society's Rowe Sanctuary near Minden, Nebraska. Thirty-one WGC members and seven spouses traveled to the sanctuary by bus to view part of the annual migration of sandhill cranes. Every year around the end of March, about 80 percent of the world's sandhill crane population converges on one eighty-mile stretch of land along the Platte River in Nebraska. Most of the cranes that stop there are sandhill cranes, but the group includes other species, such as the endangered whooping crane. After the trip, WGC's Kathy Gates submitted an article for the fall GCA *Bulletin* about "one of the greatest migrations on earth," capturing the beauty of watching the three- to four-foot-tall cranes with their six-foot wingspan swooping through the skies and landing to feast on insects, snails, and leftover grain. The WGC group observed the cranes in the late afternoon and evening of March 30 and then returned before dawn the next morning to watch the feeding frenzy. It was hard to pry the travelers away to make the simple return trip to Kansas City, nothing like the

Figure 5.15 *Left:* Sunsets in the Kansas Flint Hills continually offer a dramatic backdrop to horticultural trips. (Kilroy)

five thousand miles the cranes were traveling from their southern wintering spots in Texas, New Mexico, and Mexico to Alaska, Siberia, Alberta, and the Northwest Territories.[18]

The sandhill crane field trip included a stop on the way to the Rowe Sanctuary to visit the Lincoln, Nebraska, gallery of Michael Forsberg, a nature photographer whose photos are said to "bring the magic of nature to the minds and souls of all who see them." In 2020, Forsberg received the first GCA medal ever awarded for photography, named for J. Sherwood Chalmers. After seeing the cranes, the WGC group stopped on the way home to visit Arbor Day Farm in Nebraska City, 260 acres of natural beauty considered a national historic treasure as well as the birthplace of Arbor Day.[19]

The club made another visit to a bird sanctuary in 2016. Squaw Creek National Wildlife Refuge, located in Mound City in northwest Missouri near the border with Kansas, was established in 1935 as a resting, feeding, and breeding ground for migratory birds and wildlife. WGC members' November 18 visit allowed them to witness peak raptor and swan migration. The experience awed all in attendance, but that "bird show" paled in comparison with the club's 2014 trip.

In May 2019, Margaret Hall organized a day-long bus trip to the Village of Arrow Rock, Missouri, established in 1829 and a national historic

landmark. Besides touring historic homes, the group visited the Missouri River Bird Observatory and the Arrow Rock State Historic Site. The WGC Horticultural Book Club took an overnight trip to Arrow Rock three years later, in May 2022. Besides the aforementioned activities, the "hort" group visited area nurseries and took home bountiful seedlings to help start their summer gardens.

Figure 5.17 Other field trips focused on birds and bird migration. (Chamberlain)

Lectures Bring Awareness to the Public

The Westport Garden Club has, over the years, offered many opportunities for club members, and occasionally their guests, to learn about a variety of garden-related, horticultural, and conservation topics through lectures and presentations by experts. However, when Norma Sutherland became club president in 1977, she advocated offering educational programs to the public.

Figure 5.18 *Left:* Powell Gardens has been a longtime partner of The Westport Garden Club. (Julia Fields Jackson)

Club members were interested in the idea of public lectures because of the popularity of the presentation made to the club in the spring of 1976 by British horticulturist Frances Perry. *The Kansas City Star* not only publicized Mrs. Perry's visit but shared some of her advice: Americans assume that the Brits' climate allows for a greater latitude in plant materials, but, she advised, "You could grow a lot of our things if you would mulch more." Mrs. Perry also talked about the necessity of hedges to wall off gardens. The British, she said, prefer hedges that are both "boy and dog proof." After her visit, WGC members realized that they would be doing a public service by inviting those interested to attend lectures given by outstanding horticulturists and conservationists.[20]

According to President Norma Sutherland, "sharing Sheila MacQueen with the public" in October 1978 was the first public lecture sponsored by The Westport Garden Club. The author of thirteen books on flower arranging and a frequent speaker on BBC and for GCA clubs, MacQueen began her career in 1931 creating floral arrangements for the British royal family. She decorated Westminster Abbey for the wedding and coronation of Queen Elizabeth II and provided the floral arrangements for the Buckingham Palace wedding reception of Prince Charles and Princess Diana. MacQueen had spoken to WGC members in 1973. On her second trip to Kansas City her audience included members from several area garden clubs and professionals in the field of flower arranging. The net proceeds of the lecture amounted to $750, but "the public service rendered made it a worthwhile project," according to Norma.[21]

In 1981, WGC members decided to participate in a water study project, a topic generated by GCA. They invited Dr. Ruth Patrick, the director of the Missouri River Basin Association, to make a presentation to the club on water issues. Dr. Patrick's talk, "Water Now and in the Year 2000," had such an impact on club members that they asked her to give a public lecture, which the club sponsored. It was very well attended and convinced attendees of the importance of water conservation and efforts to improve the cleanliness of the nation's water supply. Dr. Patrick had grown up in Kansas City, attended Sunset Hill School, and, even after moving to Philadelphia, had remained a close friend of WGC founding member Barbara McGreevy. A pioneer in the study of freshwater rivers and lakes, Dr. Patrick had an eighty-year career at the Academy of Natural

Sciences. She served as an advisor on water issues to President Johnson in the 1960s and to President Reagan in the 1980s and worked with Congress on legislation that led to the Clean Water Act. Having served as an advisor to the GCA Water Project Committee, she was awarded the GCA's Francis K. Hutchinson Medal in 1977 "for her extraordinary success in gradually convincing the public and, more importantly, the industrial community that water pollution is one of the greatest dangers we face in this country." In 1982 Dr. Ruth Patrick became an honorary member of GCA. The WGC followed up Dr. Patrick's talk with a tour of the water-processing plant in Johnson County, Kansas.[22]

A Special Lecture Series in Memory of Sally Cowherd

After the successful lectures by Ruth Patrick and Sheila MacQueen, the WGC conceived the idea of holding a series of public lectures but did not have the wherewithal to do so. The opportunity came with the death of a much-beloved WGC member, Sarah "Sally" Rapelye Cowherd. Sally had served as president of the WGC from 1981 to 1983 and had supported the idea of WGC's sponsoring educational lectures. She had, in fact, made the introductory remarks before Dr. Patrick's presentation.

When Sally died in 1984, her son J. Andrew "Andy" Cowherd and her daughter Cindy Cowherd (who became a WGC member in 1999) decided to name The Westport Garden Club as one of the nonprofits to which potential gifts could be given in their mother's honor. The donations in Sally Cowherd's name were substantial enough that the club asked Andy and Cindy if they would allow The Westport Garden Club to name an educational lecture series for their mother. They warmly welcomed the idea, although they did not think the amount donated would be sufficient to sponsor more than one lecture.

The first Sally Cowherd Memorial Lecture—open to the public—was scheduled for Friday, April 25, 1986, in Curry Theatre at the Pembroke Hill School. The WGC mailed invitations to an extensive list of people, and Jo Stubbs issued a publicity statement for the media. Dr. Herbert S. Plankinton Jr., a forty-two-year-old floral designer from Longwood Gardens, near Kennett Square, Pennsylvania, with a doctorate in horticulture and a long list of impressive credentials, spoke at the first Cowherd Lecture. Dr. Plankinton had participated in and judged countless flower

Figure 5.19 The public was invited to attend a special lecture series sponsored by the WGC in memory of Sally Rapelye Cowherd.

shows, including the Chelsea Flower Show in London; lectured at arboretums and to horticulture societies; and taken his successful floral design business from Washington embassies to debutante balls and weddings in Bermuda and Hawaii.

Dr. Plankinton arrived as scheduled on Thursday afternoon, April 24, 1986, but the airline had lost both the containers and the flowers he planned to use for his presentation. While Sallie Bet Ridge Watson and Betty Goodwin ran back and forth to the airport trying to retrieve the missing items, Phoebe Hasek Bunting, Margaret "Peggy" Garner Gustin, and Norma Sutherland drove Dr. Plankinton all over town to buy more containers and flowers and to seek offerings from WGC members. Despite a wild afternoon, Dr. Plankinton was ready to begin his lecture-demonstration at 9:30 the next morning to a full house. (Fifty people were on the waiting list.) In two hours, he created a dozen designs to prove his philosophy that "flowers should unshackle the imagination." That afternoon, Dr. Plankinton gave a workshop for twenty-one WGC ladies. Those who attended the presentations found them to be very rewarding. Dr. Plankinton was wined and dined while he was in Kansas City, and he, too, considered the experience a great success.[23]

Originally WGC members had planned to make the lecture free and use $1,000 from the Cowherd Fund to pay for the event. However, the

Lecture Committee, chaired by Dorothy "Dee" Halsey Hughes, maintained that the lecture should be the first in a series, and on top of that realized that the expenses for invitations and flowers, as well as airfare, lodging, and meals for the guest lecturer, were going to total a considerable sum. Therefore, Dee Hughes and her committee decided to charge $5 for the lecture and $15 for the afternoon workshop. The tickets had imprinted on the back "We are pleased to inform you that the monies contributed in Sally Cowherd's memory will be used to sponsor an annual lecture on horticulture, flower arranging and other aspects of gardening." In the end, expenses totaled $2,446.21, exceeding income by $285.21, which came out of but did not deplete the Cowherd Fund. Thankfully, the lecture series could continue.[24]

Given the amount of work required to prepare for outside lecturers to come to Kansas City, the WGC board decided to make the lecture a biannual, rather than an annual, event. The invitation to The Westport Garden Club's second Sally Cowherd Memorial Lecture announced, "The British Are Coming! THE BRITISH ARE COMING!" On Friday, April 22, 1988, Sidney Love, a retired floral superintendent of the Royal Horticultural Society, gave a lecture and slide presentation titled "Designing a Small Garden." His wife, Daphne Love, followed with a flower-arranging demonstration. As an authorized national demonstrator for Britain's National Association of Flower Arrangement Societies, Daphne Love presented her methodology for both conditioning flowers and arranging them in "Pot Pourri of an English Garden." Following the two-hour lecture/demonstration by the Loves, again held in Curry Theatre at the Pembroke Hill School, and a luncheon for garden club members and their guests, Daphne Love presented a second workshop, limited to seventeen active WGC members. The price charged for the lecture/demonstration was $12, and it was $15 for the WGC afternoon workshop. Besides collecting these fees, WGC members sold the luncheon table decorations, and they ended up netting a total profit of $1,347, which went back into the Sally Cowherd Memorial Fund.[25]

The Loves' visit may have resulted in a "profit," but that number did not take into account the many contributions to the event, in both time and money, made by WGC members. The Loves had arrived in Kansas City on Monday, April 18, and for four days before their demonstration/lecture and two days afterward, garden club members entertained them, taking them out to eat and giving them tours of Kansas City. WGC members gathered that Sidney and Daphne Love were very surprised by the beauty and culture of Kansas City and impressed by Midwestern hospitality.

The following year, 1989, was supposed to be an off year for the Sally Cowherd Memorial Lecture Series, but when Margaret Hall, a new WGC member, proposed a program to honor her master-gardener aunts, the club embraced the idea. "Salute to Sarah and Virginia Weatherly," a program open to the public, took place on November 2, 1989, at the Junior League Headquarters, where The Westport Garden Club had taken on one of its first major projects. The club held a reception for the Weatherly sisters, followed by a program, in which *Flower and Garden* magazine editor Cort Sinnes served as master of ceremonies. Margaret presented a slide presentation depicting her aunts' garden, their lives, and their achievements. About 175 people attended this Cowherd Series salute to the Weatherly sisters.

Figure 5.20 On a small city lot, Sarah Weatherly and Virginia Weatherly developed a perennial garden worthy of being featured in the Smithsonian's Archives of American Gardens. (Smithsonian Institution, Archives of American Gardens [AAG])

Figure 5.21 The Weatherly sisters cultivated their sweet herb garden for both its utility and its beauty. (AAG)

Virginia Haynes Weatherly: Garden Celebrity

In 1997, The Westport Garden Club invited Virginia Weatherly, age eighty-seven, to become an honorary member. Virginia and her sister Sarah, who died in 1995, were well-known gardeners in the Kansas City area who found national fame late in life.

The sisters' exquisite small garden in Kansas City, Missouri—featured in numerous national magazines, a Brooklyn Botanic Garden handbook, and Rosemary Verey's book *The American Woman's Garden*—became a destination for pilgrimages by visiting gardeners such as Marco Polo Stufano, Ryan Gainey, and Bob Dash. The sisters were honored at Wave Hill, the renowned gardens in the Riverdale section of the Bronx in New York. When they were granted a Longevity of Gardeners Award, *The New York Times* featured a photograph of the sisters bedecked with laurel wreaths. In 1989 The Westport Garden Club honored the Weatherlys with a

Sally Cowherd Memorial Lecture presentation, "A Salute to Sarah and Virginia Weatherly."

The sisters traveled widely, making friends wherever they went. They corresponded and struck up a special friendship with Sir Edwin Hardy Amies, avid gardener, couturier, and fashion designer for Queen Elizabeth. He entertained the sisters in his garden for lunch.

Figure 5.22 Virginia Weatherly became an honorary member of The Westport Garden Club. (Hall)

When Sarah died, people wondered whether Virginia could cope, such was the closeness of the sisters. However, the former Hannibal, Missouri, tennis champion, who used to ride her horse to grade school, was undeterred and moved to the retirement community of Bishop Spencer Place, where she took an active part in a new garden there. A section of the garden was created by Jonathan Kemper as a replica of Virginia's previous sweet herb garden.

Although the Weatherly garden had a new owner, there was enough information about its previous incarnation for The Westport Garden Club to successfully submit it to the Smithsonian's Archives of American Gardens. When queried as to whether she missed the garden as it once had been, Virginia said, "On the contrary, gardens are meant to change with new owners and reflect their personalities." As proof of that she often enjoyed visiting her former garden. ✁

The next Sally Cowherd Memorial Lecture, "Petal Perfect," took place at the Unity Church on the Country Club Plaza and featured J. Barry Ferguson, an internationally known landscape and garden designer, horticulturist, flower arranger, lecturer, and advisor to the W. Atlee Burpee Company (an American seed and plant company founded in Pennsylvania by Washington Atlee Burpee in 1876). Despite the fact that *The Kansas City Star* had published the incorrect date for the lecture, on May 13, 1991, an audience of three hundred people enthusiastically received Ferguson, who after his lecture sold and autographed his recently published book, *Living with Flowers*. Because his lecture fee was quite high, as were his travel expenses and material costs, savvy WGC members charged twenty dollars for admission to this lecture and managed to come close to breaking even. The presentation was a great success, as was the dinner party held that evening in his honor.

In fact, Barry Ferguson was such a hit that the Project Planning Committee, headed by Margaret Hall and Dody Gates, presented the idea to the board of asking him to return in November 1992 for a fundraising event. His return trip would involve a public lecture and workshop and a separate WGC members' workshop on the following day. Billed as "Very Barry," Ferguson's return presentation had nothing to do with the Sally Cowherd Memorial Lecture Series but only with the success of his previous lecture. Eulalie Zimmer and Adele Coryell Hall planned some spectacular publicity for the "Very Barry" event, including the first page of *The Independent* and a billboard on the Southwest Trafficway.

The day of the public workshop and lecture, according to the two program coordinators, proved quite stressful for the twelve WGC members who had agreed to assist Ferguson with his presentation. They reported that he behaved like a "slave driver," uttering phrases such as "Barry wants his drink" and "Restuff all thirty topiaries." Ferguson's rudeness and arrogance caused WGC members to think they should write letters of apology to all lecture/workshop participants. However, the next day, reports from several individuals who attended the lecture/workshop relayed that they had enjoyed it. The day after, Ferguson was very agreeable during the members' workshop. According to the chairs of the Project Planning Committee, he was "charming on stage, a dictator backstage," but the end results were, at least financially, worth the pain. When Treasurer Hazel Barton reported

that the "Very Barry" event had netted the club $8,000, the stunned committee co-chairs asked her to recheck the figures: They thought they would be lucky to net $1,500 and felt sure Hazel must have miscalculated since she was at the time overwhelmed with planning her daughter's wedding. Nevertheless, Hazel's figure was correct. The fundraiser was a great financial success. Sometime afterward Barry Ferguson called and wrote to club members, apologizing for having been so "temperamental." He was, apparently, simply having a bad day. In the years since the "Very Barry" event, Ferguson has been thoughtful and kind to WGC members, including expressing his concern after the devastating 1993 Kansas City flood.[26]

Perhaps the most educational program in the Sally Cowherd Memorial Lecture Series occurred the year following "Very Barry." On November 15, 1993, Dr. Peter H. Raven, the director of the Missouri Botanical Garden and the Engelmann Professor of Botany at Washington University in St. Louis, delivered a presentation titled "Global Biology Crisis—What It Means to Us." The venue was the American Heartland Theatre at Crown Center, a location secured thanks to Adele Hall and Halls, the department store division of Hallmark Cards. Admission was free, with coffee served at 10:45 followed by Dr. Raven's lecture at 11:00. The house was packed. Sallie Bet Watson, who chaired this event, reported that "Dr. Raven was splendidly received, and all who attended came away with a very clear understanding of what the Global Biology Crisis *does* mean to us."[27]

As the funds for the Sally Cowherd Memorial Lecture Series dwindled, speakers came with less frequency. In October 1996 Shirley Dommer of the Missouri Botanical Garden delivered a presentation entitled "Bulbs for All Seasons," a timely subject that engaged the audience. In 2000, GCA honorary member Dr. Richard Lighty, a nationally known horticulturist and past head of the graduate program at Longwood Gardens in Pennsylvania, gave a presentation that would be the last of the Sally Cowherd lectures. These last three lectures—by Dr. Raven, Shirley Dommer, and Dr. Lighty—were open to the public and free of charge, which had been the hope of Cindy and Andy Cowherd from the beginning. The lecture series ended with the siblings' approval and gratitude that the Sally Cowherd Memorial Lecture Series had continued for fourteen years and with tremendous success. In 2002 the little money that remained in the Sally Cowherd Memorial

Fund and the General Memorial Fund were merged into the WGC Charitable Fund.

The cessation of the Sally Cowherd lectures did not mean the end of presentations open to the public. The WGC has partnered with Powell Gardens, with the Loose Park Garden Center, and with Deep Roots KC to bring interesting and informative experts to Kansas City to share their wisdom about conservation, horticulture, flower arranging, and garden history and design. ❧

Figure 5.23 *Right*: Margaret Hall's garden (Kilroy)

CHAPTER 6

Flower Shows Promote the Love of Gardening

WHILE SOME WGC MEMBERS BECAME immersed in issues related to conservation and the environment, others maintained that the most important part of the club's mission remained "to share the love of gardening." This included tending the plants and flowers at their homes and in the Westport Garden Room at the Nelson-Atkins Museum of Art, hosting garden tours, holding floral design and horticulture workshops, and, of course, having flower shows. The bread and butter of garden clubs, flower shows have existed in many countries and in various forms for hundreds of years. The world's most prestigious flower show, the Chelsea Flower Show in London, began as the Royal Horticultural Society's Great Spring Show in 1862 and moved to Chelsea in 1912. The largest garden and flower show in the United States is the Philadelphia Flower Show, which has been in existence for almost two hundred years. Certainly, flower shows have been important to GCA since its inception in 1913 and to The Westport Garden Club since it held its first flower show in 1955, in conjunction with its flower festival at the Nelson-Atkins Museum of Art.

As The Garden Club of America's *Flower Show and Judging Guide* explains, "The purpose of a flower show is threefold: to set standards of artistic and horticultural excellence; to broaden knowledge of horticulture, flower arrangement, conservation, and other related areas; and to share the beauty of a show with fellow club members and with the public." Each flower show

Figure 6.1 *Left:* Grant and Wendy Hockaday Burcham relandscaped their garden to include many native plants. (Kilroy)

begins with a theme and a schedule, which lists all the divisions in which a contestant can enter the show. Each of these divisions also has one or more classes, and all entries must follow the rules of the specific division and class. Flower shows are also judged, and a scale of points is established in the schedule so that participants will know how their entries will be evaluated.[1]

Much like flower arranging, the judging of flower shows is an art. In addition to being a great honor, serving as a GCA judge is a major commitment. Qualified GCA judges are recruited months in advance for every flower show presented by a member club. After ascertaining that an entry has followed the general instructions for that particular class, judges then weigh the strengths and weaknesses of the entry. Design, interpretation, creativity, distinction, and conformance are all taken into consideration in determining the winners in each class as well as the "best in show." It takes seven or more years of training to qualify as a GCA judge, and, unlike any other GCA position, once appointed, judges serve for life. They travel to and from events at their own expense and attend yearly workshops. They also are required to exhibit an arrangement of their own annually to maintain their credentials. Although judges in the nation's most prominent shows may receive small gifts and are usually provided with a meal or two, most judges fulfill their duties for the love of flower shows and in support of the mission of GCA.

As soon as the WGC joined the GCA, club members began entering and winning prizes at flower shows held at zone and annual meetings. Although WGC members readily entered shows, they found that transporting floral or horticultural designs to the shows often proved difficult. When Josephine "Josie" Jobes Giles Bunting wanted to enter a dried-flower arrangement in the 1971 Kettle Moraine Zone XI flower show being held in Nashotah, Wisconsin, WGC President Kitty Wagstaff offered to transport the arrangement on her flight to Milwaukee. However, because the arrangement was quite large, the airline required that it have its own seat and a separate plane ticket. Josie's dried flowers traveled from Kansas City to Milwaukee as Kitty's seatmate. It was worth the effort and perhaps the expense, which Josie paid, since she won a blue ribbon.[2]

The GCA Flower Show Committee promotes and provides guidance in planning flower shows, establishes and maintains standards, and encourages all GCA members to participate in both club and GCA flower shows, which differ in that club shows are in-house while GCA shows are open to the public and must include at least five divisions, one of which must be floral design or horticulture (although these requirements have relaxed over time and may vary by zone). The WGC has staged only two GCA flower shows other than those held at zone and annual meetings, but the club has recruited GCA judges for every flower show it has held.

Figure 6.2 Kitty Hall Wagstaff traveled to the 1971 zone meeting in Wisconsin with Josie Giles Bunting's dried flower arrangement in the plane seat beside her. (WGCA)

Intra-Club Flower Shows in 1979 and 1980

Back when The Westport Garden Club held the Zone XI meeting in Kansas City in 1971, WGC members were well rehearsed. The eight flower festivals that The Westport Garden Club had hosted—in 1955, 1956, 1957, 1959, 1961, 1963, 1966, and 1969—had all included club flower shows. However, by 1979, WGC members realized that they had not held a flower show for eight years, not since the 1971 Zone XI meeting. Many members had never participated in a flower show, and even those who had would need a bit of practice to do so again. Therefore, the WGC Executive Committee decided to hold their very first intra-club flower show. President Norma Sutherland reported that the idea of this simple, practice flower show caused "much grumbling and dragging of feet."

Nevertheless, it provided "a beginning from which to grow into a better, broader, undertaking next time."[3]

The following year, on July 8, 1980, WGC ladies again mounted an in-house flower show. This show, which took place at The Kansas City Country Club, also included invitations to members' spouses, who arrived after the judging to view the show, partake of cocktails, and stay for a 7:00 supper. The theme of the show was "Summer Highlights." Participants could enter three classes of flower arranging and four horticulture classes. Of course, all exhibitors' entries had to conform to the general rules for flower shows provided by The Garden Club of America.[4]

Another Flower Festival at the Nelson-Atkins

Flower shows demand much work, but they foster camaraderie, highlight talent, and generate public acclaim. Although the 1980 flower show proved entertaining and educational for WGC members, the public had not been invited to attend. In 1985, when WGC members decided after a sixteen-year break to return to the Nelson-Atkins for another flower festival, their plans also included a flower show open to the public. The festival had an Italian theme, and Frederic James, artist and husband of WGC member Diana James, designed invitations that featured the seventeen colorful Palio flags that represent the districts of Siena, Italy. As decor, Evert Asjes Jr. loaned the club twenty trees from his Rosehill Gardens nursery, including four white dogwoods, the WGC emblem tree. Ginny Hart chaired the Collector's Corner and also allowed the club to borrow several important pieces of furniture from her antique store to complement the beautiful Venetian and Tuscan pieces that came out of museum storage for the event. Boots Leiter, who would become a WGC member in 1998, worked at the time as coordinator of special events for the Nelson-Atkins. She diagrammed the festival layout and helped Festival Chair Barbara Seidlitz and her committee coordinate rentals and deliveries of plants, floral arrangements, items for sale, decorations, and lunches. WGC members also enlisted the help of thirty-four daughters and daughters-in-law. Last but not least, the club was required to have extensive liability insurance, which would have been costly had not "a very civic-minded benefactor," a member of the elusive men's auxiliary, volunteered to take care "of the premium charge for you ladies," as noted in a letter from Richard R. Callahan of Marsh and McLennan to Florence "Flip" Logan Kline.[5]

The very successful 1985 WGC Flower Festival resulted in the club issuing a check for $21,403.46 to the museum for the purpose of acquiring a large Christmas tree for Kirkwood Hall that could be used year after year. When Nelson-Atkins Director Marc Wilson wrote his thank-you note to the club, he allowed that the funds would provide not only for the cost of the tree and lighting but also for some decorations and a fine antique angel as the topper. Wilson also complimented the club for "the stylishness of the presentation," adding, "I do think everything looked quite stunning."[6]

Figure 6.3 Flower festival proceeds allowed the Nelson-Atkins to acquire a large Christmas tree for Kirkwood Hall. (WGCA)

With the labor-intensive 1985 flower festival behind them, WGC members prepared to host their third Zone XI meeting. As was typical of zone meetings, the 1987 zone meeting included a flower show. Norma Sutherland and Julia Tinsman co-chaired the show, cajoling members of the twenty clubs then in Zone XI to enter one, two, three, or even all five of the classes on the schedule. On the day of the event, Diana James stationed herself at the "passing table" and patiently inspected each entry, rattling off their botanical names "so fast it made our heads spin." An unprecedented

number of entries in both floral design and horticulture resulted in what event Co-Chairs Sallie Bet Watson and Sally Wood called "a truly magnificent affair," one that "GCA would do well to copy."[7]

Diana Hearne James: Collector and Botanist

Diana James was an elegant, reserved woman who was married to the artist Frederic James, a pupil and friend of the irascible Thomas Hart Benton. The Jameses spent summers on Martha's Vineyard in a long, ballast-stone house, where Diana kept a world-class shell library, carefully curated in special drawers built by her husband for that purpose. She was a perfectionist who did not consider her collection a hobby. Indeed, she maintained correspondence with conchologists in universities around the world. She and Fred were well known on the Vineyard for their "salons," at which the assembled wrote poems and shared and discussed classical works by authors such as Homer and Ovid.

Diana's life in Kansas City was focused on her love of horticulture. She studied botany at Kansas City University (now the University of Missouri–Kansas City) with the noted Professor John Baumgardt. Her

Figure 6.4 A woman of many talents, Diana Hearne James served as WGC president and zone representative to the GCA Horticulture Committee, while maintaining her own herbarium. (WGCA)

passion for plants led her to create her own herbarium, where she meticulously collected and conserved specimens. One memorable day while out on a walk, Diana happily climbed into a dumpster to rescue sheets of discarded botanical specimens.

She served as WGC president from 1975 to 1977 and was recognized with a Zone Horticulture Achievement Award after having served on the GCA Horticulture Committee as a zone representative and vice chair. A member of the Rare Plant Group, Diana hosted a meeting in Kansas City of that elusive and exclusive organization. Most of the members of the Rare Plant Group were in GCA clubs, and all were dedicated to preserving and propagating endangered native plants. Diana took the Rare Plant Group to view prairies on both sides of the Missouri/Kansas state line.

As they did in other endeavors, Diana and Fred worked closely together on WGC projects. Many times, the ladies of the WGC called on Fred to design posters, invitations, and flyers for events that Diana chaired, and it was he who created the dogwood design that served as the club's logo and on its needlepoint pins. ✑

1990: "A Festival of Flowers"

Three years after hosting a zone meeting, the members of the WGC put a year's worth of work into producing "A Festival of Flowers," held at Rockhurst High School on May 5, 1990. The club's most ambitious undertaking to date, the festival at Rockhurst helped celebrate the club's fortieth anniversary. Every club member held a position on at least one of the twenty different festival committees (including a "Smocks and Badges Committee"). In addition, each member donated an item for the Collector's Corner and submitted an entry for the flower exhibition, so called because it was not a judged flower show.

Invitations to the festival went out to 4,500 people, and advertisements in local publications encouraged the public to attend. Forms for preordering plants accompanied the invitations. Offerings included three different colors of hydrangeas, four colors of Rieger begonias, giant Boston ferns, and hanging baskets of fuchsias and bougainvillea in various colors. Processing these preorders turned into a nightmare, and fulfillment at the flower festival

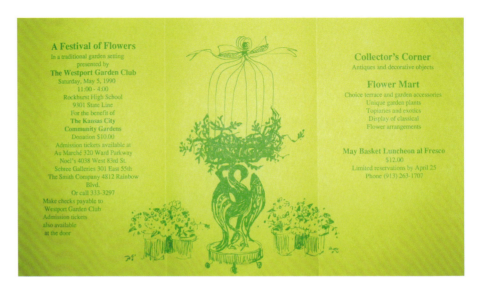

Figure 6.5 Held at Rockhurst High School in 1990, the exhibition "A Festival of Flowers" proved to be a huge undertaking.

proved to be an even greater headache. Those who had not preordered but wanted to make purchases at the event could choose from a wide variety of plants, including gardenias and hibiscus, but the preordered plants were not available to them. Some disagreements resulted, and the Flower Mart personnel felt overwhelmed trying to satisfy everyone. To provide staffing for the Flower Mart and other boutiques as well as lunch service, club members had again asked for help from daughters and daughters-in-law. Meanwhile, newer members, who had never been part of a flower festival, ran themselves ragged, according to Dody Gates, a 1990 initiate. The event left WGC members exhausted, with Julia Tinsman stating frankly at the post-festival evaluation meeting, "Our recent Flower Festival was maybe too ambitious for a club our size." She suggested that the club think about engaging in other types of ongoing projects, perhaps garden walks and small boutiques or events that emphasized conservation, which, she added, would please GCA.[8]

While garden club members recognized that "A Festival of Flowers" had indeed been a huge and at times stressful undertaking, the results seemed to justify the hard work. The WGC had designated Kansas City Community Gardens (KCCG), one of its partner organizations, as the beneficiary, and the impressive $30,000 that the festival raised enabled the KCCG to purchase a sorely needed truck, a flatbed trailer, insurance, and fuel to deliver supplies to forty active community gardens in the area. Ben Sharda, executive director of KCCG, expressed the organization's resounding appreciation to WGC for what to date was the largest single gift KCCG had ever received. In addition, the WGC used proceeds from the festival to donate $500 to Rockhurst High School to use for future plantings around its new building and $1,000 to the Rockhurst Work/Study Scholarship program. Therefore, despite the considerable effort required to orchestrate the festival, Co-Chairs Marie Bell O'Hara and Phoebe Bunting recorded in their final report: "We not only fostered good will for the Garden Club in the community, but all that hard work fostered friendships and greater companionships within the Club. Let's do it again (in about twenty years!)."[9]

After the Rockhurst festival The Westport Garden Club did indeed let a few years pass, though not twenty, without putting on a flower festival or show. Although club members continued to value the idea of holding flower shows, they would not again host flower festivals like the one held at Rockhurst and those at the Nelson-Atkins. Instead, as Julia Tinsman had suggested, they found other ways to raise money. One reason the WGC decided to cease hosting such extravaganzas was that it lacked enough members to bear the burden. Even when the motion passed in 2000 to increase total club membership from fifty to fifty-five, and then to sixty fifteen years later, the idea of another festival struck fear into the hearts of WGC members. Flower shows would suffice; no need to have them blossom into flower festivals!

Practice Makes Perfect When It Comes to Flower Shows
In October 1993, the Executive Committee decided that club members could use some practice on their flower show techniques. In answer to the directive, the WGC Flower Show Committee arranged the in-club show "Trick or Treat? A Practice Horticulture and Flower Show for WGC Members Who Are *Terrified* by the Prospect of an In-Club Show." The tongue-in-cheek title of the practice show notwithstanding, many members did express a sense of horror at the idea of an in-club show; after all, one's colleagues can be the harshest of critics! To the in-club show staged at the Linda Hall Library, club members brought branches and flowers, making

sure to follow GCA protocol barring the use of any plant materials protected by federal and/or state law as threatened or endangered. Once at the meeting, members drew lots to determine their role in the show, whether as a contestant in either the horticulture or flower-arranging division or as a judge.

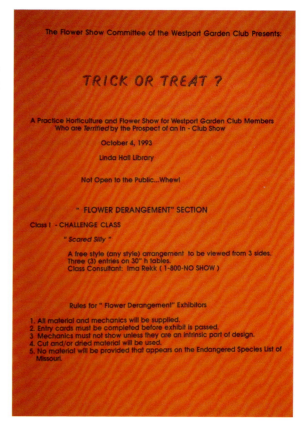

Figure 6.6 WGC members refined their techniques with an in-house practice horticulture and flower show called "Trick or Treat?"

To stay in practice for formal flower shows, "challenge shows" were sometimes staged at monthly meetings and just for fun. While club president in 1995, Margaret Hall planned a surprise Valentine's Day challenge show. Supplies issued to each contestant included white boxes, carnations, lace, doilies, ribbons, and candy hearts. Members worked in pairs to complete their arrangements. Another such challenge show took place at the Discovery Center, where members were paired and given poems to interpret through arrangements of fall leaves and other natural objects found outside.

In September 1996, to celebrate the forty-year anniversary of their club's admission to the GCA, the women of the WGC held an event they called the "Informal, Nostalgic, Non-Judged Flower Show." Each member entered one or more horticulture or flower-arranging categories, which included the opportunity to create 1) a fortieth-anniversary bouquet, 2) a ruby slippers arrangement using a red shoe in some way to celebrate the ruby anniversary, or 3) a Founders Fund arrangement, recalling the 1950s.

GCA and WGC Flower Shows Require Judges

The practice flower shows readied The Westport Garden Club for a judged club flower show. A show with the theme "Past Is Prologue" was held on Friday and Saturday, April 22 and 23, 1994, at Kingswood Manor, a senior living facility in south Kansas City, Missouri. On Friday, WGC members brought in their floral arrangements and horticultural specimens. After judging had concluded, the show opened to Kingswood Manor residents. From 3:00 to 4:30 that afternoon, so as not to interfere with the airing of the residents' favorite television program, *Jeopardy!*, WGC members served tea and encouraged the residents to tour the show. From 5:00 to 7:00, WGC friends and family previewed the show before it opened to the public the next morning. As they had since the 1960s, the members of the WGC once again donned their red smocks and wore their dogwood pins. Margaret Hall and Kathy Gates, as co-chairs, used the show booklet to explain the theme "Past Is Prologue." Citing examples of contributions The Westport Garden Club had made to the community in the past, they stated that now the club would focus on "involvement in environmental issues through support of conservation projects and education of the next generation. We strongly support recycling so that *everything old may be new again*."[10]

Although "Past Is Prologue" was a club show, not a GCA show, GCA judges from around the country came to Kansas City. GCA standards required that each class have four entries. Therefore, when someone dropped out of the recycling class at the last minute, someone else needed to fill the

Figure 6.7 *Left:* Kansas City Community Gardens benefited from the proceeds of "A Festival of Flowers." (Kilroy)

gap. In this case, Dody Gates came to the rescue. She literally put together an entry from items she collected from the trash cans and the gutter, winning second place for her display!

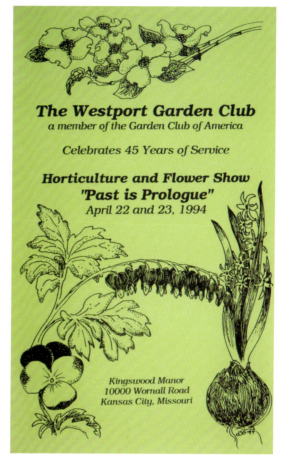

Figure 6.8 The 1994 flower show, "Past Is Prologue," demonstrated the WGC's commitment to environmental issues and recycling.

The Westport Garden Club
a member of the Garden Club of America

Celebrates 45 Years of Service

Horticulture and Flower Show
"Past is Prologue"
April 22 and 23, 1994

Kingswood Manor
10000 Wornall Road
Kansas City, Missouri

In recording the results of the "Past Is Prologue" show, President Hazel Barton said the goal had not been fundraising as much as the "sharing [of] our interests, commitment, and skills with the community." Every WGC member had participated in some way, with the show featuring twenty flower arrangements and ninety horticultural entries. The GCA judges issued "a special commendation" for the 1994 WGC flower show.[11]

In April 2002, WGC members staged another flower show, this time a GCA show with a jazz theme. Held at the Discovery Center, the show entitled "Swing into Spring" received a very good evaluation from the GCA judges, who said the show was well balanced with representations of horticulture, flower arranging, and photography, a fairly new addition to flower shows. The Westport Garden Club, in partnership with the Discovery Center, won two prestigious GCA Flower Show Awards for the "Swing into Spring" show: the Ann Lyon Crammond Award and the Marion Thompson Fuller Brown Conservation Award.

A WGC challenge show took place on March 1, 2004, with the theme "March Madness." Working in pairs, each twosome had two boxes to use as the container for an arrangement that used only the plant material provided to them in a bucket. The challenges of this in-club show and the "Swing into Spring" show helped prepare WGC members for The Garden Club of America's Annual Meeting in Kansas City in April 2005, the biggest event in WGC history. Besides a photography show and a plant exchange, top flower-arranging gurus of Zone XI took part in a floral exhibition called "New School Design."

Having fully recovered from the annual meeting, on May 14, 2008, The Westport Garden Club held another flower show, this one at the historic home of Susie Vawter. Although adjudicated by GCA judges, this show, "A Proverbial Flower Show," was a club show. It took a year to plan and was cleverly designed to use proverbs to name each of its various classes in the horticulture and flower-arranging divisions. What particularly distinguished this flower show was the class titled "Good Things Come in Small Packages," which included miniature (five inches or shorter) flower arrangements. Each arrangement was in a Lucite box that was open in the front. The miniature class was mounted in Susie's sitting room, which did not have much light, so Blair Hyde contacted a lighting designer, who created unobtrusive but very effective halogen lighting for the boxes. The arrangements and careful lighting wowed the judges!

"A Proverbial Flower Show" proved to be good practice for the GCA flower show that The Westport Garden Club hosted on May 23 and 24, 2011. Nine years had elapsed since the club had sponsored a GCA show, and this one, held at Studio B, an art studio on Main Street in Kansas City, was co-chaired by Laura Babcock Sutherland and Lyndon Gustin Chamberlain. The show was titled "Joie de Vivre: An Appreciation of All Things French." Any GCA member could participate in the six classes of photography, five

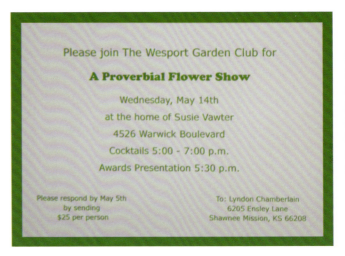

Figure 6.9 In the 2008 "Proverbial Flower Show," proverbs were used to name all the classes in the horticulture and flower-arranging divisions.

flower-arranging classes, the horticulture class, or the botanical jewelry division. Although a botanical arts division had long been an option in GCA flower shows, this was the first time for the WGC. Demanding special skill and patience, each botanical jewelry entry had to be made entirely from dried plant material and had to resemble wearable jewelry in both size and function.

The fifth division of the "Joie de Vivre" show included a conservation/education exhibit, another first for the club. The WGC's exhibit featured the Heartland Harvest Garden, an educational garden for young people that had opened in 2009 at Powell Gardens, Kansas City's botanical garden, located near Kingsville, Missouri. Three WGC members received the very prestigious Ann Lyon Crammond Award for this entry: Wendy Hockaday Burcham, Laura Lee Carkener Grace, and Laura Keller Powell.[12]

The final division of the flower show was devoted to garden history and design and featured the Horn Garden, an historic garden at Lorelei and David Gibson's estate, which was once owned by the Horns. It was documented by The Westport Garden Club in September 2007 for the Smithsonian Institution's Archives of America Gardens, a Garden Club of America project.

A club flower show entitled "Fleur-ishing Art: An American Impression" followed four years after the successful and ambitious GCA Flower Show of 2011. Co-Chairs Laura Keller Powell and Margaret "Peggy" Kline Rooney decided to stage "Fleur-ishing Art" at Commerce Bank in downtown

Figure 6.10 The "Joie de Vivre" show, held in 2011 at Studio B, an art studio on Main Street, showcased all things French. (Chamberlain)

Kansas City, Missouri, at the suggestion of Nancy Lee Kemper, wife of bank Chairman Jonathan McBride Kemper. Commerce Bank has one of the finest corporate art collections in the country, which Jonathan's great-uncle, William T. Kemper Jr., started in 1963 to provide a stimulating environment for employees, customers, and the general public. Jonathan's sister, WGC member Laura Lane Kemper Fields, who served as director of the collection from the 1980s until her untimely death in 2014, had personally overseen the acquisition and installation of the Commerce collection, not only in Kansas City but at other Commerce branches in Missouri and in Kansas, Oklahoma, Illinois, and Colorado.

Before her death, Laura Fields provided help to the "Fleur-ishing Art" co-chairs. Together they determined that each of the flower show entries in the four floral design classes and one botanical jewelry class would feature and interpret a work of art from either the bank's collection or the bank's Box Gallery, which, on a rotating basis, displayed work by local artists. In

Figure 6.11 Flower arrangements and horticultural entries allowed garden club members to share their expertise in 2011. (Chamberlain)

September 2014, the co-chairs assigned all active club members to teams according to "experience, creativity, and personalities." Sign-ups for classes happened in December, and members toured the Commerce galleries in January and February to choose an artwork on which to base their entries for the April 23–24, 2015, flower show. The "Fleur-ishing Art" show was a great success.[13]

Figure 6.12 A 2015 club flower show at Commerce Bank, titled "Fleur-ishing Art," featured floral arrangements based on works in the bank's art collection. (Chamberlain)

The Kempers, as is true of other WGC members, have a long tradition of being involved with The Westport Garden Club. Laura Kemper Fields and Nancy Lee Kemper, who are sisters-in-law, represent the third generation of Kemper women in the garden club. Enid Jackson Kemper and Ruth Rubey Kemper, daughters-in-law of William T. Kemper, had both been WGC founding members, followed in membership by Ruth Rubey Kemper's daughter-in-law, Mildred "Millie" Lane Kemper and Enid's daughter Sally Ann Kemper Wood, who served as club president from 1983 to 1985. Sally's

daughter Sharon Wood Orr is another Kemper who is a third-generation WGC member.

Once again in 2023, the WGC needed GCA judges in Kansas City as the local club planned another flower show. With the zone meeting and a flower show scheduled for September 2025, President Peggy Rooney and the WGC board decided a club flower show would be good practice for all members but especially for those thirteen active members who had never participated in a flower show before. Although this was not a GCA show, nonetheless all WGC members needed to carefully follow the GCA rules and division guidelines published in the 2021 edition of The Garden Club of America's *Flower Show and Judging Guide*. With the club gearing up much as it had for the show "Fleur-ishing Art" in 2015, every club member was urged to either enter one of the horticultural or photography classes or to pair up with another member to enter one of the floral design classes. The WGC members rallied to the call, and all in-town, active members participated, filling various roles and entering at least one division.

The show, entitled "In the Good Old Summertime," took place on May 10, 2023, at President Rooney's home. After the judging, members and spouses arrived for cocktails and a viewing. When club members assessed how much fun they had had and how much they had learned, some wondered why they did not hold flower shows more often. Those who did not wonder why were the show's organizers, Mary Levesque Harrison and Alison Ward, who had worked long and hard to bring it to fruition. The judges seemed most impressed by the peonies, the floral hats on mannequin heads in the flower-arranging division, and the amazing photographs. Photography had become an important, optional division in GCA flower shows and a part of WGC shows since the 2002 "Swing into Spring" show. The Westport Garden Club won GCA commendations for the entire class of peonies in horticulture, the "Capture the Flag" class in photography, and the "Summer's-a-Sizzling" class in floral design.

Flower Arranging:
A Competitive Sport at the KCCC Ladies' Field Day

Apart from club and GCA flower shows, for at least fifty years many WGC ladies have created floral arrangements for the traditional Ladies' Field Day event at The Kansas City Country Club (KCCC). Originally considered a

Figure 6.13 "In the Good Old Summertime" Chairs Mary Levesque Harrison and Alison Ward worked hard to bring this show to fruition in May 2023. (Kilroy)

Figure 6.14 Judges seemed most impressed by the peonies in the 2023 show. (Kilroy)

The Peonies Class

way to supply centerpieces for the luncheon tables, over the years flower arranging at the KCCC has become a competitive sport alongside tennis, golf, bridge—and more recently mahjong, fitness, and pickleball. Every participant in the Field Day events is invited to bring a floral arrangement, all of which are judged by three WGC members according to GCA standards. At the end of the day, judges announce flower-arranging prizes along with the day's other winners.

Figure 6.15 Floral arrangements, such as this one created by (from left) Dody Gates Everist, Pam Peffer, and Lorelei Gibson, have long been part of Ladies' Field Day at The Kansas City Country Club. (WGCA)

In 1987, Margaret Hall's flower arrangement won a prize at Ladies' Field Day. The "grande dames" of the WGC were so impressed that they would soon invite Margaret to join the WGC. However, before that invitation materialized in 1988, she first had to prove herself as head of the decorating committee for Ladies' Field Day. Margaret related, "I gathered a group of four others, and we were immediately told by several club grande dames that the decorations for the previous year had been skimpy and that we were to do better and have more flowers. Our budget was $50!" Considering the amount of money with which they had to work, Margaret and her committee got creative and drove around town gathering Queen Anne's lace (*Daucus carota*) and crown vetch (*Securigera varia*) from roadside ditches and vacant properties. Although their efforts did not get the women arrested, they all ended up with a case of poison ivy! To top off the experience, the lavish floral arrangement they created included a water feature, which unfortunately sprang a leak. The night before Field Day, security let the ladies back into the club to repair the damage. Manned with hair dryers, they dried the linens and patched the fountain. "The grande dames seemed pleased with the result," according to Margaret, "but I do not think that there has been such a lavish LFD since."[14]

Figure 6.16 Kristie Wolferman's arrangement won "best in show" at the 2022 Ladies' Field Day.

Many of the prizes for Ladies' Field Day flower arrangements have gone to WGC members. In June 2006, Margaret Hall served as one of the judges and, because arrangements are numbered, not named, she had no idea that five of the six winning arrangements had been prepared by WGC members: Marie Bell O'Hara, Eulalie Zimmer, Joy Laws Jones, Betty Askren Kessinger, and Nancy Stark. In 2015, Virginia "Ginger" McCord Owen won the "most creative, best use of theme" award, while Alison Ward received the prize for "best in show." Kristie Wolferman won "best in show" in 2022, and two seedling WGC members—Marianne Maurin Kilroy and Page Branton Reed—won top awards at Ladies' Field Day in 2023.

Beyond GCA Flower Shows
In 1993, two WGC members, Norma Sutherland and Susie Vawter, went beyond a GCA flower show and entered the most prestigious of all flower shows, the Royal Horticultural Society Chelsea Flower Show. According to *The Independent,*

> Norma Sutherland was asked to submit an entry representing the GCA at the Chelsea Garden [*sic*] Show in London. She was given the subject matter "A Garden of Distinction," and chose to use flowers and plant material brought to Middleton Place of Charleston, South Carolina, from England, such as dogwood, azalea, camelia, and Spanish moss. Middleton Place is the oldest landscape garden in the United States.

Norma, with Susie as her assistant, returned to Kansas City with a bronze medal![15]

Suzanne Slaughter Vawter: Keeper of Traditions
If The Westport Garden Club had created a position titled "Master of Revels," Suzanne "Susie" Vawter would surely have held it. Not only did this petite dynamo oversee the club's festivities and projects but also she was a keeper of traditions and held club members to the highest standards, always reminding them of the club's original mission.

Susie was both a horticulturist and a flower arranger, and no flowering shrub was safe from her pruners should she deem it useful for a club project. Sometimes this entailed Susie climbing a ladder to purloin branches from a neighboring church. Each fall, Susie's pollinator-friendly garden attracted hundreds of monarch butterflies, which covered her shrubs as they rested on their journey to Mexico.

Almost every club project from the 1970s on involved members working in Susie's basement—an Aladdin's cave of containers, props, tablecloths, and ribbons. Quite possibly, more club friendships were formed working in Susie's basement than anywhere else. Susie hosted flower shows, a memorable Kentucky Derby fundraiser, and club meetings in her historic home. Her legendary hospitality and generosity seemingly knew no bounds.

Figure 6.17 Susie Vawter with prominent local florist Bob Trapp. (WGCA)

An intrepid traveler who chose exotic destinations for her solo travel, Susie was also a knowledgeable collector of antiques and an accomplished guitar player.

In 1993 Susie assisted Norma Sutherland with an exhibit at the Chelsea Flower Show in England, the world's premier flower show, at which the ladies from Kansas City garnered a bronze medal. Back at home, Susie served the club in many positions, including as president. She also was the Zone XI vice chair back when the job required preparing the entire zone directory on a typewriter. Her many accolades included the GCA Club Floral Design Achievement Award and the GCA Club Medal of Merit. ❧

A few WGC members have become approved GCA judges in the fields of floral design, horticulture, and photography. For several years both Norma Sutherland and Margaret Hall served as members of a carefully selected panel of approved GCA judges who present the Bulkley Medal, which is a special award given at major flower shows. Created in 1938 by Mrs. Jonathan Bulkley of the Ridgefield Garden Club in Connecticut, the Bulkley Medal is awarded for outstanding horticulture, botany, or conservation exhibits. It is a great honor both to receive a Bulkley Medal and to be chosen as a Bulkley judge. In 2008, Norma won the GCA Zone XI Judging Achievement Award, and in 2011 Margaret served as a judge representing GCA at the World Association of Floral Artists when that show was held in Boston, its first US venue. Margaret is an approved GCA floral design judge and an approved GCA photography judge, both with emerita status. Betty Goodwin is an approved GCA horticulture judge emerita, and Lyndon Chamberlain is an active approved GCA photography judge. The WGC is proud of the many flower show and judging awards its members have received. 🍃

Flower Show Awards

GCA Major Flower Shows

Certificate of Excellence—Flower Arranging
1993 Norma Henry Sutherland (Mrs. Dwight Dierks)

Certificate of Excellence—Photography
2010 Margaret Weatherly Hall (Mrs. Thomas Bryan III)
2017 Lorelei Manning Gibson (Mrs. David William)

GCA Novice Award—Photography
2010 Helen J. Jones Lea (Mrs. Albert Robert)

GCA Flower Shows

Harriet DeWaele Puckett Creativity Award
1995 Margaret Weatherly Hall (Mrs. Thomas Bryan III)
1995 Ellison "Kelly" Brent Lambert (Mrs. Sanders Ray Jr.)
1997 Elizabeth "Betty" Kennedy Goodwin (Mrs. Frederick Merry Jr.)
1997 Sarah "Sally" Steele West (Mrs. Robert Hartley)
2011 Ellen Kirwan Porter (Mrs. Graham)

Sandra Baylor Novice Flower Arrangement Award
2002 DeSaix "DeeDee" Willson Adams
2011 Nancy Embry Thiessen (Mrs. Michael Robert)

Rosie Jones Horticulture Award
    2011 Suzanne "Susie" Slaughter Vawter

GCA Novice Award—Photography
    2011 Lyndon Gustin Chamberlain (Mrs. Richard Hunter)
    2017 Dody Phinny Gates Everist (Mrs. Brian Douglass)

GCA Novice Award—Special Divisions
    2011 Helen J. Jones Lea (Mrs. Albert Robert)—Botanical Jewelry

Dorothy Victor Munger Award
    2011 Marilyn Bartlett Hebenstreit (Mrs. James Bryant)
    2011 Nancy Lee Smith Kemper (Mrs. Jonathan McBride)

Ann Lyon Crammond Award
    2002 The Westport Garden Club in partnership with the Discovery Center
    2011 Heartland Harvest Garden Exhibit: Wendy Hockaday Burcham (Mrs. James Grant), Laura Lee Carkener Grace, Laura Keller Powell (Mrs. Richardson Kammeyer)

Marion Thompson Fuller Brown Conservation Award
    2002 The Westport Garden Club in partnership with the Discovery Center

GCA Annual Meeting Flower Shows

GCA Creativity Award—Photography
    2016 Lyndon Gustin Chamberlain (Mrs. Richard Hunter)

RHS Chelsea Flower Show

Bronze Medal
    1993 Norma Henry Sutherland (Mrs. Dwight Dierks) and Suzanne "Susie" Slaughter Vawter

Figure 6.18 *Right:* The Burchams' rock garden. (Kilroy)

CHAPTER 7

Partnerships Pay Big Dividends

WESTPORT GARDEN CLUB MEMBERS LEARNED early in their history that collaborating with other organizations helped broaden the impact of their efforts. Long-standing relationships, many of them conceived at the club's founding in 1950, continued throughout the years with city park departments and organizations such as the Nelson-Atkins Museum of Art and the Linda Hall Library.

New partnerships also emerged with Powell Gardens, Loose Park Garden Center, Kansas City Community Gardens, and Deep Roots KC, an organization founded by the WGC and originally known as the Westport Garden Club Native Plant Initiative. Also, through GCA's Partners for Plants program, WGC partnered with Kansas City WildLands, contributing funds and manpower to restore a remnant prairie in Jerry Smith Park. Thus, as a GCA club, the WGC knows well how "to share the advantages of association."[1]

That said, the term *partnership* was not in the club's vernacular until 1995, when President Hazel Barton suggested to her fellow members that they might want to consider "partnerships" when planning for future projects and activities. She had learned that "many GCA garden clubs have found this a way of undertaking significant and sizable projects." Even though Hazel realized that "our relations with the Garden Center Association, Powell Gardens, and the Kansas City Community Gardens have been greatly strengthened in recent years," she did not recognize that the WGC

Figure 7.1 *Left:* The Island Garden at Powell Gardens. (PG Archives)

already had been working with these organizations as partners to enhance the civic work of the club.[2]

In May 2020, when the WGC Long-Range Planning Committee met to study the recipients of the club's community donations, they identified three categories of partners: 1) sustaining partners, organizations with which the WGC had long relationships, which included Powell Gardens, Linda Hall Library, and Kansas City Community Gardens; 2) founding partners, organizations the WGC was instrumental in starting, such as Deep Roots KC, and 3) Garden Club of America partners, organizations that are directly associated with GCA through Partners for Plants.

An Impact on Area Cities

The WGC has had a long relationship with its local city park boards and commissioners. On at least two occasions the garden club issued statements to officials in both Kansas and Missouri to clarify the club's position on conservation and beautification. In 1967, when concerned about a proposed bridge in Mission Hills, Kansas, the club members sent out several letters, an event summarized by President Laura Kemper Carkener: "We went on record with the mayors of Kansas City, Missouri, and Mission Hills, Kansas, and the Johnson County Commissioners urging preservation of natural beauty and tasteful planning."[3]

Almost thirty years later, the WGC again "went on record" with a position paper sent to planning commissions and zoning boards on both sides of the state line. The May 1996 document, issued by the WGC Conservation

Figure 7.2 WGC members advised the Mission Hills, Kansas, Park Board on bridges and plantings. This stone bridge over Brush Creek is adjacent to The Kansas City Country Club. (Kilroy)

Committee, headed by Kathy Gates, asked city planners to think about existing trees and plantings when making decisions on new construction. Specifically, Kathy and her committee advised planners to consider how construction plans for a given property would affect irreplaceable old trees as well as trees and plantings on neighboring properties, and whether the plans allotted sufficient green space. "As stewards of your city," the WGC position paper read, "we request that you include these criteria in your decision making. . . . Trees and plantings contribute immeasurably to the environment and beauty of older neighborhoods. When contemplating approval of plans for new building, we ask that you not ignore these valuable assets." And finally: "Our recommendations are in keeping with the guidelines of our national organization, The Garden Club of America."[4]

The Kansas City, Missouri, Park Department helped the WGC plan some of its first civic projects, while the club has contributed several unsolicited opinions to the Park Department on landscaping and beautification. At the club's January 1981 meeting, Millie Kemper moved that

the WGC write a letter to the department, by then called the Parks and Recreation Department, expressing their opposition to the permanent installation of reflective pylons (a project by sculptor Dale Eldred) on the median of Brush Creek Boulevard south of the Nelson-Atkins Museum of Art. The motion passed unanimously. Adele Hall, wife of Donald J. Hall, chairman of Hallmark Cards, volunteered to hand deliver the letter to the Parks and Recreation Department meeting on January 6th. Ultimately the Eldred installation did not happen. In her 1981 year-end report, President Barbara Seidlitz stated that the WGC had "played a part in keeping the city beautiful."[5]

WGC members spoke out when they noted a civic issue, and they also recognized the work of others. In 2013, the WGC proposed Louis Larrick Ward and Adelaide Cobb Ward for the GCA Club Civic Improvement Commendation. The Wards' foundation had funded projects in Kansas City, Missouri, which "benefitted parks throughout the city and served as an example for others to follow to encourage the continued maintenance, renewal, and enhancement of the green spaces that keep our city beautiful." Adelaide "Addie" Ward attended the annual WGC meeting to receive the commendation on behalf of herself and her late husband along with her sons Tom and Scott and Scott's wife, Alison Ward, a WGC member.[6]

Figure 7.3 Surrounded by family members, Adelaide Ward took part in a ceremonial tree planting at Loose Park in Kansas City, Missouri. (Photo courtesy of Alison Ward)

In some ways, club members have played an even more vital role in Mission Hills, Kansas, than they have in Kansas City, Missouri. WGC members have served on the Mission Hills Beautification Committee (now called the Mission Hills Park Board) since the garden club's inception. Mission Hills is a unique residential area envisioned in 1912 by developer J. C. Nichols, who wanted to create a garden community of country estates. He hired the landscape architect Sid J. Hare and his son S. Herbert Hare to plat three hundred acres of wooded land just west of the Missouri state line. Following the natural contours of the rolling terrain, Hare & Hare laid out curving streets and constructed rustic stone bridges across the area's meandering creeks. Along the wide drives connecting many large and some smaller estates, Hare & Hare established fifty-seven "parklets," which created settings for the display of classical statuary, fountains, and urns that J. C. Nichols had collected in Europe.

While J. C. Nichols took measures to help ensure the upkeep and beautification of the area, it was his son Miller who implored the WGC to take an active role in overseeing the city's landscaping needs. Miller Nichols knew of The Westport Garden Club through his sister, Nicky Allen, a founding member. Thus, in 1950, when the City of Mission Hills founded its Beautification Committee (the same year as the founding of the WGC), Miller Nichols appointed to the committee five WGC members who resided in Mission Hills. Today the Mission Hills Park Board has a full-time arborist, a maintenance firm, and an expert who annually checks the fountains and statuary. Three teams of volunteers, representing the city's three homes associations (Mission Hills, Indian Hills, and Tomahawk Road), monitor their areas for landscape issues, take notes on what bulbs and annuals to plant, and make recommendations for improvements, in consultation with a representative from Rosehill Gardens. Although now the Park Board requires only one WGC member, there are usually more on the teams. In 2023, WGC members on the Mission Hills Park Board included Chair Lorelei Manning Gibson, Susan Chadwick Pierson, Mary Ann Powell, Britton Franke Norden, and Sharon Orr.[7]

Figure 7.4 *Right:* Hare & Hare, landscape architects, established fifty-seven parklets in Mission Hills, which created settings for statuary, fountains, urns, and flowers. (Kilroy)

At one time Lorelei Gibson, who has served as the Mission Hills Park Board chair for the last twenty-five or more years, reported on Mission Hills beautification at WGC meetings. However, those reports ended with the September 2011 meeting, at which a member pointed out that not everyone in the WGC lived in Mission Hills. The club, therefore, realized the need to recognize that Kansas City is a bi-state area with numerous municipalities in which WGC members live—including Kansas City, Missouri, and many cities in Kansas in addition to Mission Hills: Fairway, Lawrence, Leawood, Mission, Mission Woods, Overland Park, Prairie Village, Westwood, and Westwood Hills. In fact, only about a quarter of WGC members live in Mission Hills.

However, even those members who do not live in Mission Hills care about keeping this unique, premier residential city beautiful. When GCA meetings and garden tours have taken out-of-towners through the area, the club has asked the City of Mission Hills to step up its beautification plan. The city planted 250 Mount Hood narcissus before the 1987 zone meeting and tulips and pansies before the 2005 Annual Meeting.[8]

Museum Connection Diminishes over the Years

Even after the cessation of flower festivals at the Nelson-Atkins, the connection between the museum and the WGC remained important. In 1988, thirty-eight years after the club started its first museum project, President Sallie Bet Watson reported that "the Nelson-Atkins Museum consumes the major part of our efforts and funds." WGC members handled the maintenance of the Westport Garden Room, installed in 1976, until the museum decided to repurpose the space in 2003. The WGC continued to arrange flowers for special exhibits and festive occasions through the 1980s. For the museum's fiftieth anniversary in December 1983, the WGC provided thirty flower arrangements.[9]

A Christmas tree the WGC purchased with funds from the 1985 flower festival gave rise to the Nelson-Atkins's annual Christmas tree lighting ceremony, attended by hundreds of people. A decade later, when this tree had fallen into disrepair, the WGC again staged a fundraising event, called

Figure 7.5 *Left:* Delegates attending the 2005 Annual Meeting of GCA in Kansas City enjoyed seeing the tulips that the City of Mission Hills had planted. (Kilroy)

"Art of Flowers," which raised $20,000 for a new tree. In thanking WGC members, Director of Development Michael S. Churchman explained, "The tree has become a tradition which is enjoyed by countless visitors. We are indebted to The Westport Garden Club for creating and continuing to support this celebration."[10]

Figure 7.6 Garden club members handled the maintenance of the Westport Garden Room, installed in 1976, until the museum repurposed the space in 2003. (WGCA)

Having established the Junior Gallery and Creative Arts Center in 1960, the WGC supported the gallery's Art in Nature classes through the 1980s and invited children in the classes to visit members' gardens to sketch ideas for art projects. However, in 1981 the instructors at the Junior Gallery decided to use the bus fare money donated by the WGC to transport children to the zoo instead of a member's garden. In 1986 they curtailed trips to members' gardens altogether, using the funds to instead purchase items "that supplemented our resource center and dealt directly with the Art in Nature theme," according to Creative Arts Coordinator

Penny Selle. New Art in Nature programs continued to be introduced, concluding in 1986 with the successful and well-attended Wildflower Studies.[11]

Beginning in 1980, through the Spang Memorial Fund, WGC members began providing the museum with floral arrangements for Thanksgiving and Mother's Day and paperwhites in celebration of Chinese New Year. Two brothers, Thomas Green "Tom" Spang and Joseph Peter "Peter" Spang III, conceived the idea of the Spang Memorial to honor their mother, Gwendolyn Green Spang, who had died when the boys were very young. Their father, Joseph Peter Spang Jr., had met his wife when his employer, Swift & Company meat packing, sent him from Chicago to Kansas City. After his wife's death, Spang moved to Massachusetts with Tom and Peter. However, the boys spent summers in Kansas City with their Grandmother Green.

When, as adults, they returned to Kansas City for a conference, they realized that no trace of their mother's family remained in her hometown. They decided to ask the Nelson-Atkins Museum of Art for permission to place a plaque and to display floral arrangements there in honor of their mother and her family. When they talked to Ross E. Taggart, then curator of decorative arts at the museum, he suggested they ask The Westport Garden Club to take on the project. Upon hearing from the Spang brothers, WGC President Sally Cowherd consulted her board, and the club agreed to help them honor their mother's memory. Barbara Seidlitz and Sallie Bet Watson created the first such floral arrangement in 1980. The initial donation by the brothers was for $2,000, then raised to $3,000. After Tom Spang died in 2009, his widow, Caroline, continued to support the project; she made a final donation of $10,000. In 2022, the balance of the Spang Memorial Fund, which is held in an account at the museum, was $8,200. WGC members continue to use their floral design expertise to honor the Spang family, placing arrangements three times a year in an area designated by the museum. No one seems to remember the plaque meant to honor Gwendolyn Green Spang.[12]

Apart from the arrangements for the Spang Memorial, all that remains of the WGC's many contributions to the Nelson-Atkins Museum of Art is a plaque in recognition of the former Westport Garden Room.

Figure 7.7 Every year, through the Spang Memorial Fund, garden club members prepare a floral arrangement at the Nelson-Atkins for Thanksgiving. (WGCA)

An Enduring History with Linda Hall Library

Linda Hall Library is the longest continuing partner of The Westport Garden Club. For seventy-five years, apart from when the library was closed for remodeling in 2005–2006, WGC members have, as mentioned, met there, and library personnel have assisted with meetings, displays, and the harboring of the club's archives, located in a large closet in the bowels of the library.

Over the years, the club has given the library many gifts, including rare books, wrought iron gates, a garden figure and pool, plantings, and trees for the arboretum: crabapple trees (*Malus*) in memory of library trustee Frank W. Bartlett, a grove of twelve Scots pine trees (*Pinus sylvestris* 'Argentea') in memory of nurseryman Evert Asjes Jr., as well as various trees planted in memory of WGC members. In any given year, the WGC plants a tree at

Linda Hall Library, Powell Gardens, or Kansas City Community Gardens memorializing members the club has recently lost.

In 2009–2010, WGC helped fund the creation of a brochure and map of the Linda Hall Library Arboretum as well as the development of a database, signage, and QR codes for the more than four hundred trees on the library's grounds, including, at that time, twelve trees identified as Greater Kansas City Champion Trees. A $5,000 gift to Linda Hall Library from the WGC in 2013 included funds for general maintenance of the library and for the development of a viburnum collection on the northwest side of the arboretum grounds. A bench near the viburnums is dedicated to club member Vivian Pew Foster. Other benches on the north side of the library were given "in loving memory" of Katherine Histed Foster, a founding member, and Sallie Bet Watson, who served as club president from 1987 to 1989. In 2005, Sallie Bet was awarded posthumously the WGC "Above

Figure 7.8 In memory of Evert Asjes Jr., the WGC planted a grove of twelve Scots pine trees at Linda Hall Library. (LHLA)

and Beyond Award," recognizing extraordinary service. Each year the club also contributes an annual stipend to the library in appreciation.[13]

Periodically, the Linda Hall Library Arboretum arborist has given WGC members guided tours of the library's grounds, most recently in 2011 and 2022, when she reminded members of the importance of the tree peonies, viburnum, and champion trees.

Enhancing the Beauty of Loose Park

Jacob L. Loose Park, the third largest park in Kansas City, Missouri, has historical significance. In the early days of settlement, William Bent, a fur trader and wagon train leader, owned the seventy-five acres that now comprise the park as well as additional acreage. His property was a major site of the Civil War Battle of Westport in which Union forces routed the Confederates. After Bent died in 1871, Seth Ward bought the land from Bent's estate and, in 1896, agreed to lease the property for one dollar per year plus taxes to a group of men who had incorporated The Kansas City Country Club and wanted to establish a golf club. However, the club had to relocate in 1926, when Ella Loose bought the seventy-five acres from the estate of Seth Ward's son Hugh Ward. Mrs. Loose then gave the property to the city in 1927 to use as a public park and named it Jacob L. Loose Park for her late husband, founder of the Loose-Wiles Biscuit Company, which produced Sunshine Biscuits.[14]

Once it was established, members of the Kansas City Rose Society saw Loose Park as a place to realize their dream of a rose garden open to the public. In 1931 the Rose Society, under the leadership of Laura Conyers Smith, established the garden, which in 1965 was renamed for her as the Laura Conyers Smith Municipal Rose Garden. Designed by eminent landscape architect S. Herbert Hare, the original garden contained 120 rosebushes; today the garden includes 3,000 rosebushes of nearly 130 varieties, all of them maintained through a partnership of the Kansas City Parks and Recreation Department and the Rose Society. The Westport Garden Club joined this partnership when in 2000 the Rose Society embarked on a three-phase plan to restore the Rose Garden. The WGC donated $100 to Phase I in 2001 to help rebuild the fountain and $6,000 as part of Phase III, which involved the enhancement of the West Garden and included the installation of a bench in memory of Sallie Bet Watson. The Rose Garden, restored to respect Hare's original design, reopened in 2006 in time to celebrate its seventy-fifth anniversary. The World Federation of Rose Societies awarded the garden a Garden of Excellence award in 2018, and the Kansas City Rose Society received the 2018 GCA Club Horticulture Commendation, proposed by the WGC. Besides having contributed to the restoration of the Rose Garden, several WGC members belong to the Rose Society, and a few deceased members have roses in the garden donated in their memory.[15]

In 1980 the Kansas City Parks and Recreation Department designated the collection of trees in Loose Park as an "arboretum" and named it after Stanley R. McLane, the former head landscaper for the J. C. Nichols Company. To celebrate Arbor Day in 1981, the Parks Department planted an additional sixty-seven trees of various varieties. The WGC helped purchase aluminum plaques to mark each of the new trees with its common and scientific names. Two years later, in 1983, the Arbor Day Foundation and the Missouri Department of Conservation named Kansas City a "Tree City," a designation of the Tree City USA program that the city sustained for many years and that requires a municipality to have a viable tree-management plan and program.[16]

Tree Camp

In 2007 the Kansas City Parks and Recreation Department asked the Loose Park Garden Center to initiate "an educational program at the park" for children attending the department's community centers' summer programs. The parks department hired Rita Shapiro to direct the program and help secure its funding. Garden Center Association President and WGC member Virginia "Ginny" Bedford McCanse assumed the task of creating the program. Her inspiration came from two sources: 1) a map of the trees in the park that philanthropist Mary Atterbury had drawn for her grandchildren; and 2) a prairie field guide made by Wendy Paulson ("The Nature Lady" and 2017 GCA medalist) that Ginny had seen at the GCA Zone XI meeting held in Barrington, Illinois.[17]

The WGC contributed funds for the production of *Explorers' Field Guide: Exploring and Observing Loose Park's Arboretum*, published by the Garden Center Association in cooperation with the Board of Parks and Recreation Commissioners of Kansas City, Missouri. Ginny McCanse wrote the guide

while future WGC member Jo Meyer Missildine designed the map and the logo for the book. Two members of the Sierra Club—Eileen McManus and Dave Patton—also contributed to the guide, along with Molly Fusselman, a horticulture educator at the University of Missouri Extension. Judy K. Bellemere created and donated the book's artwork. The Garden Center Association made the field guides available for individuals and families to pick up and use to explore the park, but the guide also served Tree Camp, the educational program envisioned by the Parks and Recreation director and created by Ginny McCanse.

Figure 7.9 Ginny McCanse wrote the field guide and Jo Meyer Missildine created the map and logo to introduce Tree Camp participants and families to Loose Park. The Tree Exploration Course was suitable for very young children.

Prior to the first session of Tree Camp, held the third week of June 2008, volunteers from the Garden Center Association and the WGC—including Blair Hyde, Jo Missildine, Gina Miller, Ginny McCanse, Jill Kathryn Stewart Bunting, and future WGC member Kristie Wolferman—held a fundraiser/

Figure 7.10 *Right:* WGC members contributed to the restoration of the Rose Garden at Loose Park. (Kilroy)

practice session during which they learned how to use the guide. They also equipped each of the three hundred backpacks purchased by the WGC for the campers' use, packing them with the necessary supplies, including the guide, a compass, a magnifier, binoculars, a clipboard, a pencil, and a tape measure, to navigate and record observations in the field. During the week-long Tree Camp, each day about thirty-five children, ages six to fourteen, attended a 10:00 a.m. to 3:30 p.m. session, with the Parks Department providing lunch. Three teachers along with many volunteers staffed a crafts center and led campers through 1) the Lake Course, which required students to use the map in the field guide to find fifteen special trees; and 2) the Champion Tree/Compass Course, in which campers needed to follow the guide to learn how to measure trees and use a compass to find the ten trees in Loose Park identified as champion trees, the largest trees of their kind in Kansas City. These trees included a swamp chestnut oak (*Quercus michauxii*), a littleleaf linden (*Tilia cordata*), a blue ash (*Fraxinus quadrangulata*), a Douglas fir (*Pseudotsuga menziesii*), an American sycamore (*Platanus occidentalis*), an Amur cork (*Phellodendron amurense*), a yellow buckeye (*Aesculus flava*), a weeping white pine (*Pinus strobus* 'Pendula'), a bigleaf linden (*Tilia platyphyllos*), and a red hickory (*Carya ovalis*).

Figure 7.12 During Tree Camp, Jo Missildine instructed children on how to use the field guide to measure trees and find the champions. (WGCA)

Figure 7.11 Campers could use a magnifier in their Tree Camp backpacks to get a closer observation of the bark on a tree. (WGCA)

Tree Camp was such a success that the next summer, in 2009, the WGC purchased three hundred more backpacks and again provided volunteer staffers. After that, the Parks and Recreation Department took over the program, holding its last Tree Camp in 2010; nonetheless, the Garden Center continued to make the *Explorers' Field Guide* available to visitors. The Westport Garden Club helped introduce young people and their families to the Jacob L. Loose Park and the Stanley R. McLane Arboretum, provided tools and volunteer staffers, and thereby set the example of how to best explore and appreciate the park and the natural world.[18]

Figure 7.13 Tree Camp participants tromped through Loose Park carrying backpacks filled with necessary supplies—guide, compass, magnifier, binoculars, clipboard, pencil, and tape measure. (WGCA)

Figure 7.14 Ginny McCanse, who spearheaded the club's Partners for Plants project, sorted seeds collected at Jerry Smith Park. (WGCA)

Virginia Bedford McCanse: Perpetual Motion Machine

Virginia "Ginny" McCanse is a bundle of energy. She rides her bicycle for miles on end; she has hiked the 567-mile Colorado Trail (average elevation: 10,300 feet) and climbed mountains in Patagonia, South America.

However, nothing prepared her for her job at the GCA's 2005 Annual Meeting in Kansas City. She had been a WGC member for only three months when the annual meeting chairs asked her to co-ordinate and oversee the premeeting visit by the GCA Horticulture Committee. This obligation included arranging the group's visitation of nineteen private gardens in two days, orchestrating two dinners (with a tornado alert sounded during one of them), and seeing to the needs of the Plant Exchange Committee. In her thank-you note to Ginny, Cinder Dowling, chair of the GCA Horticulture Committee, wrote, "When The Westport Garden Club accepted you as a member, they should have been bowing down in appreciation."

Ginny was energized rather than exhausted by her "baptism by fire" and feels that her annual meeting experience helped her gain an early understanding of the GCA. She went on to lead WGC's Partners for Plants invasive plant eradication project at the Jerry Smith Park remnant prairie. As president of the Loose Park Garden Center Association (now Gardeners Connect), she partnered with Powell Gardens and WGC to offer a series of garden lectures. She instituted Tree Camp for children and wrote a field guide for the park based on one she had seen at a zone meeting. She also co-chaired a major WGC fundraiser, "Entertaining Gardens." Not surprisingly, she was awarded both the GCA Club Horticulture Award and the GCA Club Appreciation Award, plus a special club award for going "above and beyond."

Ginny now resides in Tucson, Arizona, where she grows her cactus garden solely from cuttings. When not engaged in a competitive game of pickleball, she leads botanical tours of the nearby canyons. ❧

Lecture Series Educates the Public

The Loose Park Garden Center was built in 1957 to house a horticulture library and to provide space for meetings and horticultural exhibits. Over the years, the WGC has made contributions to the Garden Center to enable a lecture series and to provide books for the library (including a book given each year in honor of the WGC president). Many WGC members have belonged to the Loose Park Garden Center; ten members belonged during the time Paget Gates Higgins served as WGC president (2003–2005). Several WGC members have held positions on the Garden Center's board, and both Ginny McCanse and Jo Missildine have served as board chair. For her leadership, Ginny received the Arbor Day Honor Award presented by the Kansas City, Missouri, Parks and Recreation Department on April 12, 2009. WGC members gathered in Loose Park to witness the planting of a Cornelian cherry dogwood (*Cornus mas*) in Ginny's honor. Seven years later, in 2016, Parks and Recreation also planted a tree in Loose Park in honor of Jo Missildine.[19]

When the Missouri Department of Conservation established the Anita B. Gorman Discovery Center in 2005, the Loose Park Garden Center lectures moved to the Discovery Center's auditorium. At the time, the WGC donated $5,000 to the Loose Park Garden Center Lecture Series to promote and facilitate excellent lectures and programs. The WGC Executive Committee stated, "It seems like a natural fit for the WGC to support and eventually co-sponsor some of the lectures as part of our outreach to the community." Certainly, a healthy synergy existed between the WGC, the Loose Park Garden Center, and the Discovery Center.[20]

The WGC's establishment of the Loose Park Monarch Garden in 2016 and the two shade gardens in the park in 2018, a planned renovation of the azalea garden there in 2025, and ongoing maintenance of these gardens would further strengthen the ties between the WGC and Loose Park.

Years of Friendship with Powell Gardens

Powell Gardens, a botanical garden in Kingsville, Missouri, about thirty miles east of Kansas City, sits on 970 acres of "lush, rolling hills and windswept meadows," more than 175 acres of which are open to the public for "education, exploration, and recreation." Visitors from across the country come to view the approximately 6,000 varieties of plants and take advantage of seasonal festivals, while annually more than 15,000 young people take part in Powell Gardens' educational programs and nature activities. A nonprofit organization, Powell Gardens enjoys the generous support of foundations, corporations, and private individuals.[21]

The Westport Garden Club's association with Powell Gardens began in 1988, when WGC member Sarah "Sally" Steele West served on the first Powell Gardens Board of Directors. Another connection with the WGC was made when the Gardens hired Eric Tschanz as executive director in 1989. Tschanz came from the San Antonio, Texas, Botanical Garden, where he had worked closely with WGC member Jean McDonald Deacy, who had strong ties with the San Antonio garden and whose family foundation was based there.

Clarence Halsell Holmes, Jean Holmes McDonald Deacy, Virginia McDonald Miller, Laura Woods Hammond: A Family Quartet

Although family connections among The Westport Garden Club's membership are far from unusual, only one family can claim a four-generation membership from great-grandmother to great-granddaughter, the Holmes/McDonald/Miller/Hammond family.

Clarence Halsell Holmes

The matriarch, Clarence "Clare" Holmes, was born in Vinita, Oklahoma, a wide space in the road in what was then Indian Territory. Her father was a rancher, one responsible for developing tiny Vinita into a prosperous cattlemen's town. The family pet was a wolf, and Clare and her siblings ranged freely on the vast grasslands on which their family's home and ranch sat. When Clare and her two sisters reached high school age, their father sent them to a "finishing school" in New England. The Halsell family moved to Kansas City when Clare was twenty-one. There she met Jay Holmes, a lawyer specializing in real estate. Later Jay would work for Clare's father in land development, primarily large swathes of ranchland in Texas, Kansas, and Oklahoma.

Figure 7.15 *Right:* Powell Gardens sits on 970 acres of "lush, rolling hills and windswept meadows." (PG Archives)

Although she looked fondly upon her ranching background and never lost her love of cattle and horses, Clare immersed herself in the social life of Kansas City. Before long she became joint master of hounds at the Mission Valley Hunt Club, rode in the American Royal Horse Show, and joined The Westport Garden Club as an early member. Her home boasted an extensive garden designed by Hare & Hare landscape architects and done in an Italianate style, with multiple terraces, waterfalls, and fountains. When Clare's granddaughter Virginia "Gina" Miller was visiting the Philbrook Museum of Art in Tulsa, Oklahoma, many years later, she experienced a strong sense of déjà vu when viewing the gardens there. The Hare & Hare design of the Philbrook garden struck her as being nearly identical to her grandmother's garden in Kansas City.

Figure 7.16 Jean Holmes McDonald Deacy, here in her youth, and her mother, Clare Halsell Holmes, an early WGC member, shared a love of horses similar to their love of the land. (WGCA)

Jean Holmes McDonald Deacy

Clare's daughter Jean inherited her mother's love of horses and cattle, but she had a different type of upbringing. Jean graduated from the Sunset Hill School in Kansas City and went on to attend Smith College, where she majored in economics. She married William "Bill" McDonald in 1941. Bill was the true gardener in the family, but the lure of the Kansas prairie and cattle ranching soon captured him. Jean was warm, vivacious, and social. Her leadership abilities meant that she served as the president of many organizations, including the Kansas City, Missouri, Junior League and The Westport Garden Club. Jean also sat on multiple boards, including that of the Lady Bird Johnson Wildflower Center in Austin, Texas. Jean and Bill gardened at their home in Kansas City as well as at their residence in La Jolla, California. Guests recalled memorable evenings at the latter during which the fragrance of their hosts' night-blooming cereus cacti filled the air.

Virginia McDonald Miller

Jean's daughter Gina inherited the gardening gene, but in her case it entailed a passion for the natural rather than the cultivated world. Gina, too, attended the Sunset Hill School. As a young wife and mother, Gina lived briefly in Illinois. Her daughter Laura remembers Gina's food garden, and that she let Laura and her brother decide what plants to grow—popcorn, peanuts, and rhubarb being perennially popular choices.

Returning to Kansas City from Illinois, Gina began accompanying her father to informative cattle meetings in San Antonio. There she discovered what would end up being her lifelong passion of cattle breeding—this in the early days of crossbreeding cattle for specific traits. In addition, Gina began to learn about the native prairie plants and grasses as they related to the health of livestock, identifying both nutritious and harmful plants. She wanted to develop the ability to identify all species encompassed by a loop of rope she had thrown anywhere on the ground. Because of this interest, Gina gravitated to three WGC projects: Deep Roots KC, the Loose Park Monarch Garden, and Partners for Plants.

Field trips to Gina's Mashed O Ranch in the Flint Hills of Kansas always proved to be a special treat for WGC members. The ranch is especially memorable for those 2005 Annual Meeting attendees lucky enough to go on the post-meeting trip there.

Gina's good stewardship of her ranch resulted in her being recognized with both the GCA Club and GCA Zone XI Conservation Awards. Additionally, she received the Morris County, Kansas, Grassland Award and a recognition from the Missouri Prairie Foundation.

Figure 7.17 Gina McDonald Miller, Jean Deacy's daughter, inherited the gardening gene and developed a keen interest in native prairie plants. (WGCA)

Figure 7.18 Laura Woods Hammond, Gina Miller's daughter and fourth-generation WGC member, enjoys participating in the wreath project. (Kilroy)

Laura Woods Hammond

The fourth member of this WGC quartet is Gina's daughter Laura Hammond. Laura joined the WGC when her daughter Sophie was in college. Laura's interest in conservation was jump-started when Kathy Gates, who was preparing to leave town, asked her to look after a monarch chrysalis during her absence. Laura had the requisite milkweed, and soon the pupa hatched and became a monarch butterfly whom Laura named Cedric. Laura was convinced that Cedric recognized her since he seemed to greet her whenever she stepped outside. She later read that butterflies may retain memories from their chrysalis stage.

Laura has immersed herself in the garden club, serving as club Conservation Committee chair, joining the Horticulture book club, conducting research in the club archives, and entering a club flower show in which she and her mother won the Novice Award. Time will tell whether Laura's daughter Sophie will become the fifth generation of the family to join WGC. ❧

Since the establishment of Powell Gardens, the WGC has encouraged its members to purchase memberships there, and the WGC community liaison reports regularly on activities at the Gardens. Marjorie Powell Allen and her brother, George E. Powell Jr., conceived the idea of Powell Gardens, and members of the Powell family have also served as conduits between the Gardens and the WGC. Wendy Jarman Powell became a WGC member in 1996; Laura Powell did so in 2010; and Mary Ann Powell joined in 2015, becoming the WGC president in 2023.[22]

The year after Eric Tschanz came to Powell Gardens, he hired Spencer Crews as the first director of horticulture. Crews generously shared his expertise with WGC members, making numerous presentations to the club. He left Powell Gardens in 1996 to become director of the Lauritzen Botanical Gardens in Omaha, Nebraska. In 2010 Crews became a GCA honorary member from Zone XI, and in 2013 he received the Zone XI Civic Improvement Commendation for his creation and continuing development of Lauritzen Gardens (where the Shirley Meneice Horticulture Conference was held in 2017). Crews, a member of Omaha's Loveland Garden Club, returned to Kansas City in 2023 to speak at a WGC meeting about his experience as the first director of Lauritzen.

In 1996, Director Eric Tschanz hired horticulturist Alan Branhagen to replace Crews at Powell Gardens. Both Branhagen and Tschanz led tours of the Gardens for WGC members, conducted tree walks in parks and cemeteries, raised sapling trees in the Gardens' greenhouses for an all-club GCA project, held workshops to help WGC with horticultural entries for flower shows, served as program speakers at club meetings, and made themselves available to club members with horticultural questions.

Figure 7.19 Eric Tschanz, former executive director of Powell Gardens, has led tours of parks and cemeteries for WGC members. (WGCA)

Alan Branhagen remained the director of horticulture at Powell Gardens for twenty years. One of the many projects he oversaw was the Legacy Tree Project of Kansas City, a program designed to find and propagate the area's champion and historical trees. The Kansas City area champion tree list, which had been started in 1955 by Stanley B. McLane and updated by Chuck Brasher, had been monitored and maintained by Powell Gardens since 2012. Branhagen used this list to seek out champion trees—both native and ornamental—and propagate them. In October 2013, he gave a talk to the WGC, "Our Legacy of Trees," and after his presentation took club members on a walking tour in Loose Park to view some of the historic and champion trees. In 2014, Wendy Powell wrote an article in GCA's publication *The Real Dirt* that explained the Legacy Tree Project of Kansas City.

As it turned out, many of these historic trees were found living in old cemeteries, so Branhagen also conducted several tree walks through cemeteries. In November 2015, he led WGC members through Forest Hill Calvary Cemetery, established in 1888. At one time, Forest Hill was known nationally for its trees, especially for its fine collection of Missouri natives and its beautiful sugar maples (*Acer saccharum*). The following year, Branhagen led a tour through Elmwood Cemetery, sighting both historic and champion trees. The fact that Elmwood Cemetery, which opened in 1872, has the feel of a park is not an accident, for it was laid out by the renowned landscape architect George Kessler, who worked in collaboration with August Meyer, the first president of the Kansas City, Missouri, Parks Board. The 43-acre Elmwood Cemetery has 337 trees, several of historic significance, including the cemetery's signature elm.

Alan Branhagen also helped the WGC with the design of the Loose Park Monarch Garden and with the refurbishment of a garden in Joplin, Missouri, after the town was devastated by a tornado. In recognition of his expertise as a naturalist specializing in birds, butterflies, and botany, the WGC successfully proposed Branhagen for a GCA Zone XI Horticulture Commendation in 2013. Even after Branhagen left Powell Gardens to become the director of operations and horticulture at the Minnesota Landscape Arboretum at the University of Minnesota, he maintained contact with WGC members. He returned to Kansas City to help the garden club develop a plan to revive and rehabilitate a neglected azalea garden in

Loose Park. On July 15, 2023, Branhagen assumed a new position as the executive director of the Natural Land Institute in Rockford, Illinois, but he continued to offer his assistance to the WGC for the azalea project.

Figure 7.20 Alan Branhagen, former horticulture director at Powell Gardens, led a tour through Elmwood Cemetery, sighting both historic and champion trees. (WGCA)

Meanwhile, Eric Tschanz, retired from Powell Gardens in 2017, became an honorary WGC member in 2006 and remains active in the club. His expertise has been invaluable in developing and maintaining the WGC native plant gardens at Loose Park.[23]

The Westport Garden Club continually supports Powell Gardens as a sustaining partner, donating money every year since 1995. In 2005, the Gardens received some $10,000 from the WGC, seed money for the development of its Heartland Harvest Garden, which broke ground on October 1 of that year. In 2009 the Heartland Harvest Garden opened as the nation's largest edible landscape, featuring twelve acres of edibles and companion plants. The garden was designed for youth education, with the rationale that "children who understand where food comes from are more apt to eat fresh, healthier foods and engage in healthy lifestyle choices." The completion of the Heartland Harvest Garden brought national attention to Powell Gardens in "seed to plate" issues and sustainable agriculture. In 2010 the

WGC made another large donation ($10,985.07) to Powell Gardens, and the club's "Joie de Vivre" flower show in 2011 featured a conservation/education exhibit from the Heartland Harvest Garden.[24]

The WGC's "Entertaining Gardens" fundraiser, held in 2013, benefited Powell Gardens' Good to Grow educational program, providing funds ($36,500) to subsidize the program and purchase furniture for the children's classroom in the education building. The WGC event invited the public to tour members' gardens and to shop at a boutique that featured garden supplies as well as potted succulents, terrariums, and fairy gardens, all planted by WGC members. In subsequent years, the WGC has made annual donations to Powell Gardens.[25]

As a means of educating the public, Powell Gardens and The Westport Garden Club have partnered to bring in a number of speakers, including the 2013 GCA Margaret Douglas Medal recipient, Doug Tallamy. Powell

Entertaining Gardens

The Westport Garden Club

Figure 7.21 The "Entertaining Gardens" fundraiser, held in 2013, benefited the Good to Grow program at Powell Gardens.

Gardens staff members have made many presentations to the WGC, have helped with several of the club's horticultural projects, and have enjoyed many years of friendship with the WGC.

Kansas City Community Gardens Flourishes in the Urban Core

Kansas City Community Gardens (KCCG) is a nonprofit organization whose mission is "to empower and inspire low-income households, community groups, and schools in the Kansas City metropolitan area to grow their own vegetables and fruit." KCCG offers low-cost seeds and plant varieties that thrive in the local climate, sponsors workshops, and provides technical support to get gardens growing in the Kansas City area.[26]

KCCG began as the Metropolitan Lutheran Ministry's Community Garden Project in 1979 but was incorporated in 1985 as KCCG with headquarters in the former Old Ballpark Community Garden at 22nd and Brooklyn. Several WGC members volunteered at KCCG, and, in 1989, alongside other organizations, the WGC began working to help KCCG find a new location that would allow for larger administrative offices and garden space. That location turned out to be in a Kansas City, Missouri, park. On March 20, 2001, a ground-breaking ceremony was held at Swope Park, with many WGC members and friends present.

Soon after the ground-breaking, WGC members started a new project with KCCG—the Beanstalk Children's Garden, "a unique learning environment where children are invited to see, smell, touch, and taste growing plants." To enable the WGC to provide the seed money for the children's garden, Lorelei Gibson and her husband, David, hosted a benefit at their home on September 22, 2001. The cocktail buffet and garden tour, open to WGC members, spouses, and friends, raised enough money to jump-start the installation of the Beanstalk Children's Garden in 2004. WGC members received invitations to the grand opening on June 21, 2005, along with invitations to work at the garden in a variety of volunteer positions! The Beanstalk Children's Garden incorporates seven gardens—the Fruit Garden and Demonstration Orchard; the Curiosity Garden, which is full of unusual plants; the Water Garden; the Vegetable Garden; the Seed and Grain Garden; the Herb Garden; and the Insectary Garden, which attracts, feeds, and shelters beneficial insects. Besides hosting many visitors, the Beanstalk

Figure 7.22 The entryway to the Beanstalk Children's Garden at KCCG invites children into a different kind of learning environment. The WGC provided the seed money for the garden, which held its grand opening in 2005. (Kilroy)

Children's Garden holds educational and scout badge workshops; 3,800 children attended these programs in 2022.[27]

In 2006 the WGC members voted to organize the Family Fall Festival at KCCG for the purpose of introducing new, prospective donors to the organization. Even though "our club has supported KCCG with volunteer time and funds since 1989," WGC President Sally West wrote in her annual report for 2006–2007, "our goal was to make the Kansas City community more aware of the meaningful services the gardens provide to low-income families and school groups. By providing instruction, tools, seeds, and plots of land, KCCG enables people to plant, identify insects, and understand the reasons for making the right environmental decisions." The first fall festival attracted many people who had known nothing about the work of

Figure 7.23 *Right:* Corn and other edibles, as well as flowers, grow at Kansas City Community Gardens. (Kilroy)

KCCG and who left better informed and also willing to help grow the organization by donating time and money. A WGC all-hands-on-deck effort led to the success of the festival. Each member had four responsibilities for the event: 1) to purchase a ten-dollar family ticket to the event, 2) to invite four or more other families to attend, 3) to volunteer a minimum of two hours at the event, and 4) to bring at least two dozen cookies to pass out to the children. KCCG continues to hold its Family Fall Festival every September. Although club members still serve as volunteers and provide cookies, KCCG is now fully in charge of the event.[28]

At the WGC Annual Meeting, held on June 6, 2023, at KCCG, Director Ben Sharda reported on the long history of collaboration between the two organizations and the major monetary contributions and time commitments made by garden club members. While Sharda spoke on the extent to which WGC has helped KCCG over the years, it bears mention that he and the KCCG staff have added much to the benefit of the city and to the education of WGC members. In 2014, Sharda received the GCA Club Civic Improvement Commendation, and in 2015, he became an honorary member of The Westport Garden Club.[29]

One of the programs KCCG operates is The Giving Grove, the mission of which is "to provide healthy calories, strengthen community, and improve the urban environment through a nationwide network of sustainable little orchards." The Giving Grove supports neighborhood volunteers who plant and care for fruit trees, nut trees, and berry brambles. What began as a single orchard in 2013 grew in a few years to more than one hundred orchards in neighborhoods that had faced decades of environmental and health inequities. Inspired by its string of successes in Kansas City, The Giving Grove expanded nationwide in 2019. Today more than 800 volunteer orchard stewards manage some 480 orchards in thirteen cities across the United States: Atlanta, Auburn, Cincinnati, Dallas, Denver, Detroit, Kansas City, Louisville, Memphis, Omaha, South Bend, St. Louis, and Seattle. Besides reinvigorating urban spaces, the groves in 2022 provided 64.5 million servings of fresh fruit and nuts. In 2019 The Giving Grove received a GCA Club Horticulture Commendation, proposed by the WGC.[30]

The continuing growth of KCCG is impressive. In 2022, staff and volunteers cultivated 167,644 vegetable and herb transplants, prepared 45,861 seed packets, and delivered 66 educational workshops, thus providing help to 3,179 home gardeners, 315 community gardens, and 231 neighborhood orchards. Also during 2022, KCCG worked alongside partners to start 31 new gardens and 20 new orchards and to expand 38 successful sites. These city gardens harvested 1.2 million pounds of fruits and vegetables in 2022, benefiting 42,438 households. In addition, KCCG supports 223 school gardens and 58 school orchards, engaging over 19,000 students. The WGC is proud to consider KCCG its partner in helping to improve the environment and feed the city.[31]

The "Deep Roots" of Native Plant Initiatives

In February 2014, The Westport Garden Club invited Doug Ladd, conservation director for The Nature Conservancy for the state of Missouri, to present a program at the club's monthly meeting. He had last spoken to the club in 1986, a presentation that inspired the WGC to take a trip to the Grassland Heritage Foundation. Ladd's 2014 talk again focused on the prairie, which he described as a precious resource and one of the most threatened and least conserved of the world's ecosystems. WGC members took Ladd's presentation as a call to action. Honorary WGC member Robert J. "Bob" Berkebile, an environmentalist and founding partner of BNIM Architects, issued the club members a challenge "to become pollinators and advocates for native plantings." That spring, WGC President Nancy Lee Kemper asked Jo Missildine and Kathy Gates to organize a Native Plant Task Force with the mission "to explore the possibilities to promote native plants."[32]

Kathy Gates took the leadership role, becoming the one "crazy lady" determined enough to see that the club would make a difference. Twenty members joined the WGC Native Plant Initiative, experimenting with native plants in their gardens and researching what existing organizations did to benefit native landscapes. They organized a September 2014 trip to The Nature Conservancy's Dunn Ranch Prairie to better understand what The Nature Conservancy was doing to restore more than a thousand acres of the remnant prairie there.[33]

The WGC also partnered with Powell Gardens to provide educational opportunities for the public to learn more about native plants. They organized a presentation by the entomologist, ecologist, and conservationist Doug Tallamy from the University of Delaware, an honorary member of GCA and a Margaret Douglas medalist, as well as by Roy Diblik, a garden designer, champion of native plants, author, and the owner of Northwind Perennial Farm in Burlington, Wisconsin. In 2018 the GCA bestowed an honorary membership upon Diblik and in 2023 awarded him with a Medal of Honor for his "outstanding service to horticulture."[34]

With Kathy Gates at the helm, and supported by Bob Berkebile, the WGC decided that the opportunity existed to create a larger organization than The Westport Garden Club Native Plant Initiative. They personally contacted elected officials, architects, landscapers, and representatives from nonprofit organizations, conservation agencies, and local businesses to determine whether any of them had an interest in working together toward promoting the use of native plants, supporting ecological balance, and seeking solutions for clean air and water. Representatives from ten municipalities, four park districts, and groups from the private sector convened. They voted unanimously to collaborate, resulting in the formation of the Kansas City Native Plant Initiative (KCNPI).[35]

KCNPI, later renamed Deep Roots KC, became a collective-impact organization of some seventy multisector partners, all of whom shared the vision of fostering beautiful, native landscapes connecting communities in which "nature and people thrive." The organization stated its mission as "to encourage the appreciation, conservation, and use of native plants in the heartland through educating, collaborating, and facilitating the planting of regenerative native landscapes that are essential for a healthy planet." Deep Roots KC touts the slogan "What you plant matters," shorthand for "What you plant matters: for clean water and air, for pollinators and our food supply, for butterflies and birds, for the future of our climate, for human physical and mental health."[36]

WGC's dedication to the concept of increasing the use of native plant landscapes led to two club projects—the installation of the Loose Park Monarch Garden in 2016 and two native plant shade gardens in the park in 2018. In addition, beginning in 2016, the WGC initiated a native plant sale, which served as a club fundraiser for several years. Interest in planting native species ran high in the Kansas City area, as indicated by the fact that at the KCNPI native plant sale, held June 3, 2017, in Shawnee Mission Park, volunteers sold a thousand plants in under five hours.[37]

Westport Garden Club members continue to support Deep Roots KC and to stay informed about the many programs offered by the organization,

Figure 7.24 Deep Roots KC's slogan is "What you plant matters." The organization, started by the WGC, encourages the use of native plants. (WGCA)

including online and in-person learning opportunities, native plant sales, and conferences. Deep Roots held Plan It Native, the first regional native landscape conference, on September 18–20, 2019. The WGC donated $10,000 to underwrite the keynote speaker, Florence Williams, author of *Nature Fix*, who described in her presentation the many benefits of nature for humans. Because of COVID-19, the annual Plan It Native conference was conducted virtually for the next four years, but in 2024 returned to an in-person event. The Deep Roots website offers extensive resources for gardeners of all abilities for planning, planting, and managing native gardens. Its e-newsletter, *The Pollinator*, is an additional monthly source of information delivered to thousands of KC-area inboxes.

In 2017, Kathy Gates received the GCA Club Conservation Award for her part in the organization of the WGC Native Plant Initiative and her absolute dedication to seeing it expand into Deep Roots KC, an important regional collaborative. Kathy then received the GCA Zone XI Conservation Award in 2018. GCA also recognized Bob Berkebile's contributions to this initiative and to conservation in general, first with a GCA Zone XI Conservation Commendation in 2021 and then with the Cynthia Pratt Laughlin Medal, presented to him at the 2023 GCA Annual Meeting "for his outstanding conservation advocacy and helping to form the US Green Building Council and its LEED rating system."[38]

Kathy Garrett Gates: A Voice for Native Plants

Thinking of her days growing up in Chagrin Falls, Ohio, Kathy Gates cannot remember a time when she wasn't gardening. As a child, she used to create sandbox villages replete with mosses and plant material. When she moved to Kansas City with her husband, it was perhaps inevitable that she would join The Westport Garden Club. Her mother-in-law was a member, as are two sisters-in-law and a niece. At first, the club's formality gave Kathy pause, especially being listed in the club's directory as Mrs. Kirkland Hayes Gates and referred to in the club's minutes, which were always read aloud at meetings, as Mrs. Kirkland Gates. (Formal titles remain in the directory.) A bit of a rebel, when she served as the club's recording secretary, she daringly used her fellow members' first names in the minutes, and no one even noticed.

Figure 7.25 Kathy Garrett Gates strongly advocates for native plants. She grows them in her own garden and promotes their use to encourage pollinators and preserve our ecosystem. (Kilroy)

In those early years of her membership, Kathy enjoyed all the club's offerings and even co-chaired a flower show. Her favorite club-related activities included field trips and horticultural projects, and she found attending the National Affairs and Legislation Committee meetings inspiring. Nevertheless, she spent most of her time tending to her young children and her career as director of the Interclub Tennis League.

Then came a club meeting in February 2014 that changed her life. The speaker was Doug Ladd, a noted prairie expert and the conservation director of The Nature Conservancy of Missouri. Ladd told the group, "The prairie is the most endangered, rare, and unique ecosystem on the planet." That single sentence served as a call to action for Kathy, igniting a passion that eventually led to the 2014 founding of the Kansas City Native Plant Initiative (now known as Deep Roots KC). A collaborative effort with members representing numerous municipalities in both Kansas and Missouri, this group dedicates itself to the vision of a healthy ecosystem and the creation of more native landscapes in the Kansas City area. Kathy is quick to credit the WGC for supporting this effort, claiming, "All it needed was a champion." Deep Roots now sponsors an annual conference with attendees from all over the world and provides activities and resources to the region.

Kathy also helped the WGC found its monarch habitat garden at Loose Park after she became alarmed by the plight of the migratory pollinators. Kathy has been recognized with both the GCA Club and Zone XI Conservation Awards as well as the GCA Club Horticulture Award.

Does Kathy herself have a favorite native plant? Yes, it is the bright orange wildflower known as the orange, or hoary, puccoon (*Lithospermum canescens*). ◈

Saving the Prairie at Jerry Smith Park

In 1976 Jerry Smith, longtime Kansas City civic leader and philanthropist, gave 360 acres of land at 139th and Prospect to the City of Kansas City. This property provides a southern anchor for the Centennial Boulevard— forty-three miles of boulevards and parkways that extend from Kansas City International Airport in the north to this land, Jerry Smith Park, in the south, connecting most of the city's major parks. From 1977 until 1984, Kansas City's Parks and Recreation Department maintained Jerry Smith Park for educational purposes as a working farm with fruit trees, vegetable gardens, livestock, and hayfields, all managed by Jewel and Phyllis Loveland. When the Lovelands retired, the park was used by hikers and nature lovers who enjoyed exploring its rolling hills, wooded forests, the three-acre lake, and forty acres of natural prairie.

In 2009, Ginny McCanse and then WGC President Ann Parker North Readey attended Prairie Days, a Kansas City Parks and Recreation event held at Jerry Smith Park. The festive event included live music, plein air painting, crafts, a covered wagon serving buffalo burgers, and tours. There Ginny and Ann met Larry Rizzo, a biologist with the Missouri Department of Conservation. He took them out into the midst of tall grasses to see an area of native prairie, land that, Rizzo explained, had never been plowed and that held species of flora and fauna that could only have grown in undisturbed soil. Rizzo informed the ladies that Kansas City Parks and Recreation and the Missouri Department of Conservation had a partnership with Kansas City WildLands, an organization that coordinates volunteers dedicated to the protection of biodiversity through hands-on restoration and management of natural communities. Jerry Smith Park needed maintenance and restoration to protect its rare native species and to eradicate invasive honeysuckle and other noxious plants. Looking out at the gorgeous prairie, Ann said, "I'd like our club to have a GCA Partners for Plants project, and this could be it." Ginny recalls that she did not know what that meant, but she consulted the GCA website and discovered that the mission of The Westport Garden Club dovetailed perfectly with that of Partners for Plants. Moreover, what became known as the Jerry Smith Park Project would further the WGC's goals of fostering native plants and pollinators.[39]

Actually, some members of the WGC had become acquainted with Partners for Plants (P4P) four years earlier, back in 2005, when the WGC had asked GCA for suggestions for the use of the $16,214.26 in extra funds from the annual meeting that GCA had held in Kansas City. GCA had suggested P4P. The idea of forming P4P had originated at a GCA Conservation and National Affairs and Legislation Committee Conference in 1991 and was launched in 1992 with the stated purpose "to facilitate hands-on projects between local GCA clubs and land managers on federal, state, local, and other significant public lands." Each P4P project is unique, and the GCA provides funding to any proposed project that meets the P4P mission statement and guidelines. The Kansas City WildLands/Jerry Smith Park Project fit the criteria.

"All it took from our club," Ginny McCanse explained, was for "someone (me it turned out) to be the Project Sponsor and apply for grant funds." In 2009, WGC became one of the first clubs in Zone XI to apply for a P4P project grant and receive funding. At the time, $3,000 was available per year, and from 2009 to 2015 these grant funds went toward the eradication of the very invasive *Sericea lespedeza* (an Asian legume known as the "Plague on the Prairie") and bush honeysuckle (*Lonicera tatarica*) in Jerry Smith Park. As part of the project, WGC volunteers joined others from around the city at the park to identify and mark rare native plant communities in the spring, collect seeds in the fall, and separate seeds from their husks during the winter months. To this day, every trip to Jerry Smith Park provides an extraordinary learning experience in a beautiful setting.[40]

Figure 7.26 Ann Bunting Milton, Laura Babcock Sutherland, Ginny McCanse, Diana Jackson Shand Kline, and Wendy Powell worked at Jerry Smith Park, eradicating invasive plants and gathering seeds. (Photo courtesy of Ginny McCanse)

A botanical survey of Jerry Smith Park in 2011 provided scientific proof of the biological integrity and quality of the park's remnant prairie. In 2016, P4P grant money helped fund a species study at the park. The study resulted in the identification of more than 200 species of plants, and the 2017 seed collection revealed even further diversity. In 2022, volunteers collected 206 pounds of seeds with a retail value estimated at $32,000. The harvest included the collection of 136 native, local genotype species, including 12 high-conservation-value species unavailable commercially anywhere else. The seed team, many of whom were WGC members, contributed 850 hours collecting, processing, and hand-broadcasting seeds back onto

Figure 7.27 *Left:* Long-horned bees (genus *Mellissodes*) are very active from July through September at Jerry Smith Park. One lights on a prairie aster (*Symphyotrichum turbinellum*). (McCanse)

remnant prairie—a job of priceless value to local ecosystems and one that brought the participating club members great satisfaction.[41]

Besides the link to P4P, the Jerry Smith Park Project put the WGC in partnership with Kansas City WildLands, the Missouri Department of Conservation, and the Kansas City Parks and Recreation Department. Ginny McCanse took the lead on the project. When she moved to Arizona, Ann Bunting Milton took over as the project coordinator, followed by Gina Miller. Throughout, the project mission statement has remained consistent: "to assist in the restoration of the only known, unplowed prairie left in Jackson County, Missouri, by making an inventory of plant species and bees and by propagating and planting native plants." Through their involvement in the restoration of the Jerry Smith Native Prairie Park, WGC members can claim a role as stewards of the last remaining natural prairie in the Greater Kansas City area.[42]

Figure 7.28 Prairie blazing star (*Liatris pycnostachya*) thrives on the remnant prairie at Jerry Smith Park. (McCanse)

Collaboration with so many organizations has increased the productivity and influence of The Westport Garden Club locally, regionally, and sometimes even nationally. To support the dynamic partnerships and various projects the club has taken on, the women of the WGC have had to find creative ways to consistently raise the needed funds, something the next chapter will address. ❧

Figure 7.29 *Right:* Plantings form a quatrefoil at Powell Gardens. (PG Archives)

CHAPTER 8

Supporting Our Partners and Funding Our Projects

SUCCESSFUL FUNDRAISING IS VITAL TO virtually all of a nonprofit's endeavors, which certainly holds true for The Westport Garden Club. Although WGC members pay dues annually, the revenue derived from dues covers only the club's direct operating costs and its GCA dues. In the very early days of the WGC, after voting on which projects to approve, club members anted up the money to pay for them. That way of proceeding ended after the first year. Thereafter the WGC held fundraisers, some of which, as we have seen, were annual events held to support partners and sponsor civic projects to further the club's mission of sharing the love of gardening "through education, action, and creative expression in the fields of horticulture and conservation."[1]

Early projects, as discussed above, involved relandscaping the Westport Triangle, helping fund the restoration of the Wornall House and its grounds, and making major improvements to the Junior Gallery and Creative Arts Center at the Nelson-Atkins. Later projects included designing the Westport Garden Room; creating a thatched cottage for the Kansas City Flower, Lawn, and Garden show; starting a hortitherapy program at the Florence Crittenton Center; planning and planting the playground at the Children's Center; participating in the Tulips on Troost Project, the Joplin Project, the Kansas City Museum garden project, the 2013 Centennial Tree Project, and the Olmsted 200 Project; and establishing the Monarch Garden, two shade gardens, and an azalea garden at Loose Park.

Figure 8.1 *Left:* Kathy Gates's garden. (Kilroy)

The Westport Garden Room: A Salute to the Bicentennial

Just months before the WGC's twenty-fifth anniversary celebration in 1975, GCA announced a recommendation that each of the 181 GCA member clubs embark on a special project to celebrate the nation's bicentennial in 1976. WGC President Diana James conceived the idea of honoring the occasion by creating a garden room at the Nelson-Atkins Museum of Art. Her ties to the museum extended far beyond her WGC involvement; she and Karen "K.D." Bunting (mother-in-law of WGC member Jill Kathryn Stewart Bunting) had opened the museum's Sales and Rental Gallery in 1958. Moreover, she knew that the garden room project would appeal to WGC members. The club's Executive Committee took the concept to the museum's board of trustees. They quickly approved the idea of creating a garden room in space that was then a hallway leading out to the Pierson Sculpture Garden. Representatives from the WGC and the museum began working with an architect.

Fundraising for the Westport Garden Room began at the home of Jean McDonald, with an "Affair of Chance" cocktail party. Doris "Dorry" Mead Gates—mother of Paget Gates Higgins, mother-in-law of Kathy Gates and Dody Gates, and grandmother of Newell Gates Brookfield (all WGC members)—held another event at her E. G. Gallery and gave a percentage of the sales for the evening to the garden room fund. One last benefit happened in May 1975, when club members and their spouses enjoyed a picnic dinner in the Pierson Sculpture Garden at the museum. Last, but far from least, two generous anonymous donors contributed $40,000, which brought the room to reality.[2]

On April 30, 1976, "A Garden Celebration," which was open to the public, inaugurated the Westport Garden Room. Flowering plants, topiaries, and arrangements decorated the new room, and club members offered a sales boutique of garden accessories and plants, most of which had been raised in Millie Hoover's greenhouse. The Garden Club of Cleveland lent a huge bicentennial eagle for the opening. In the Atkins Auditorium, several WGC members gave talks on gardening and showed a Williamsburg film, *A Colonial Naturalist*.

In writing his last annual report as the director of the Nelson-Atkins in 1976, Laurence Sickman noted, "The Westport Garden Club made another of its really important contributions to the beauty of the Gallery with the construction of a delightful garden room, some 30 by 36 feet, leading directly onto the Pierson Sculpture Garden," through two sets of French doors. Interspersed among the plants in the garden room were four eighteenth-century figures given to the museum by the garden club in 1966.[3]

The Westport Garden Room also housed a beautiful fountain, a gift to the museum from the George L. Gordon family and dedicated in memory of WGC founding member and Nelson-Atkins patron Jane Hemingway Gordon. Originally, the centerpiece of the fountain was a sculpture by Anna Hyatt Huntington. When the museum remodeled and recast the room as the Aquila room in 2003, Margi Conrads, the Samuel Sosland Curator of American Art, decided that the existing sculpture was inappropriate for the new design. Instead, she selected a bronze fountain figure by Harriet Whitney Frishmuth called *Joy of the Waters*, given to the museum in 1996 as a bequest from Elizabeth Abernathy Hull. The choice was serendipitous; Margi had no idea about Miss Hull's connection to the WGC. The room donated by the WGC no longer exists, but a plaque identifies the Aquila room as the original Westport Garden Room.[4]

When she was WGC president in 1977, Norma Sutherland reported that the club's greatest joy "was in seeing our Garden Room at the Nelson Gallery, conceived in 1975 and born in 1976, mature in 1977." With the new stringent protocol against allowing live plants and flowers in the museum's galleries, the garden room provided WGC members a place to display their talents. When the museum hosted the *Sacred Circles* exhibit in 1977, WGC members outdid themselves with a lavish display of lilies, hydrangeas, and Boston daisies (*Argyranthemum frutescens*). In 1979, when Claude Monet's

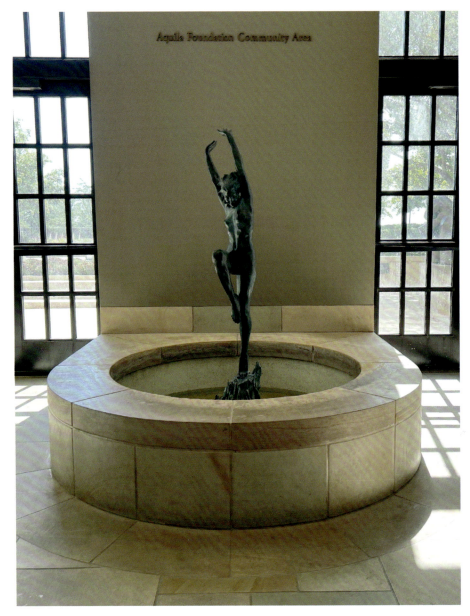

Figure 8.2 The fountain in the original Westport Garden Room sits now in what is called the Aquila room of the Ford Learning Center. It features a sculpture given to the museum by Elizabeth Abernathy Hull titled *Joy of the Waters*, by Harriet Whitney Frishmuth, American (1880–1980). (Image courtesy of Nelson-Atkins Media Services, Jeff Nightingale)

Water Lilies triptych went on display, WGC members made special floral arrangements for the garden room. In addition, in the loan gallery leading into the Monet exhibit, they created a French allée, a straight passageway lined on either side with white apple trees planted in stunning square tubs and surrounded by white Japanese irises. WGC President Barbara Seidlitz recorded in her annual report that club members agreed this display was "one of our most successful civic efforts."[5]

A rotating maintenance crew of WGC members tended the permanent plants in the garden room and provided seasonal blooming plants. For more than twenty-five years, maintaining the Westport Garden Room was an ongoing and major commitment for club members. Over time, the beauty of the space began to fade along with enthusiasm for its upkeep. In the final days of the room, DeeDee Adams remembers watering the plants and then taking a tongue lashing from Director Marc Wilson, who followed a stream of water from a leaking planter into galleries housing precious art. The re-purposing of the room by the museum was a bittersweet end to the project, but it actually came as a relief to both the garden club and the museum.[6]

Building an Irish Cottage, Thatched Roof and All

When asked in 1977 to participate in the municipal Flower, Lawn, and Garden Show scheduled to be held in Bartle Hall February 11–19, 1978, club members, who considered the invitation an honor, voted a resounding yes. Members thought their display would aid in public relations and demonstrate their goodwill in supporting a large civic endeavor. Back in March of 1963, the club had had a booth at the Flower and Garden Exposition in which it exhibited horticultural and floral design pieces, for display only, including Flora Barton's "Ivy Eagle," winner of an award of merit at the Chicago World Flower and Garden Show in 1962. And again in 1965, WGC members had prepared an entry for the Flower and Garden Exposition held at the Municipal Auditorium, creating a small patio exhibit and selling pink geraniums and grape ivy, which won the display contest for area clubs. In February 1966, the floral-decorated WGC booth at the Kansas City Garden Show, set up by Vivian Foster, Maxine Goodwin, Hattie Byers, Laura Kemper Carkener, and Georgette O'Brien, had provided information about GCA and the WGC. However, in 1977, the club was given not just a booth but a twelve- by fourteen-foot space to fill.

Wanting to create something to please the thousands who would attend the show, garden club members decided, for some reason long forgotten, to create an authentic Irish thatched-roof cottage. They enlisted help from the men's auxiliary. Jack O'Hara drew the blueprints, and Sutherland Lumber Company constructed the cottage, thanks to Herman Sutherland. The thatching, however, was done by Club President Norma Sutherland, Marie Bell O'Hara, Libby Adams, and Millie Hoover, who twenty-seven years before had served as the WGC's first president. Other club members participated in building a low rock wall around the cottage and positioning pots of flowers to look as if they had been planted. The February scheduling of the garden show limited horticultural options. Therefore, the previous December, Millie Hoover, who had a greenhouse, purchased strips of dormant sod from a turf farm. She watered, fertilized, and trimmed the grass in her greenhouse with scissors to make it an even, lush green "lawn." Other members forced white hyacinths and white alyssum, and a local nursery forced a redbud tree, but the club wished for primula and pansies, which were not available. Then "an indispensable husband came to our rescue," explained Project Chair Libby Adams. He had a day's business in Houston and took some members along to purchase flowers. (While there, they had a lovely lunch with Helen Anderson from The Garden Club of Houston.)

Figure 8.3 When the WGC was given a booth at the 1978 Kansas City Flower, Lawn, and Garden Show, members decided, for unknown reasons, to design an Irish cottage and garden, admired here by Norma Henry Sutherland. (WGCA)

Every day of the Flower, Lawn, and Garden Show, club members took shifts to care for the grass and flowers and to explain the purpose of The Westport Garden Club. The thatched cottage, however, had demanded such tremendous effort that it proved to be The Westport Garden Club's last undertaking of the kind. Still, as Libby Adams reported, "We had marvelous weeks of fun, and everyone fell in love with our Irish cottage."[7]

Hortitherapy Benefits
WGC Members as well as Crittenton Girls

All gardeners know the joy that comes from digging in the earth, planting seeds, watching seedlings come up and thrive, and then enjoying the beauty and the harvest that result from their labor. However, until WGC members visited the Menninger Clinic in Topeka, Kansas, they had not realized that gardening can have real therapeutic value. In November 1978, WGC members took a full-day field trip to learn about the Menninger hortitherapy program. At the clinic they heard talks given by three horticultural therapists who traced the development of horticultural therapy and its use and value in treating patients with various nervous and/or mental disorders. As a result, four determined WGC members set in motion a hortitherapy project for Florence Crittenton girls.

The Florence Crittenton Home is a long–standing institution in Kansas City and in other cities as well. The concept began in 1882, when Charles N. Crittenton's four-year-old daughter, Florence, died of scarlet fever. Wracked with grief and seeking a way to channel his sense of loss, Crittenton decided he could make a difference by devoting his time and wealth for the betterment of a needy group—unwed, abandoned, and/or homeless women who lacked medical care and perhaps housing for themselves and their infants. He opened the first Florence Crittenton Home in New York City in 1883 to provide shelter for such women and their children. Then, with co-founder Dr. Kate Waller Barrett, Crittenton started the National Florence Crittenton Mission, which received a federal charter under President William McKinley. Crittenton purchased a railroad car and took "The Good News Train" coast to coast, opening Florence Crittenton Homes in many cities across the country. Eventually, he established sixty-five homes in the United States as well as twelve abroad. The "Crittenton Movement" proved so successful that the National Association of Florence Crittenton Agencies carried on Crittenton's legacy for decades after his death. The National Crittenton Foundation still exists today.[8]

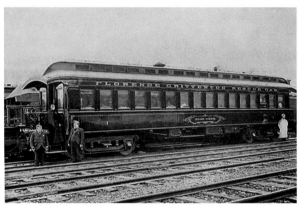

Figure 8.4 Charles N. Crittenton took "The Good News Train" on a journey across the country to establish Florence Crittenton Homes in city after city, arriving in Kansas City in 1896.

In 1896 "The Good News Train" arrived in Kansas City, and with the support of fourteen local businessmen, Crittenton launched a Florence Crittenton Home at Fourth and Main. The home provided shelter, food, and adequate medical care for unwed mothers and their children. In 1925, a similar institution called the Florence Home was established for single black mothers by Elizabeth Bruce Crogman. Local philanthropist William Volker, the founder of the University of Kansas City (now the University of Missouri–Kansas City), provided funding. The Florence Home and the Florence Crittenton Home merged in the early 1970s. The merger of the two homes required more space, and in 1978 the new home, called the Florence Crittenton Center, moved into a facility at 10918 Elm Street.

When the center moved to its new location, Crittenton staff called on The Westport Garden Club to get some help with landscaping. The garden club voted to use $1,000 to plant the small front courtyard. With an additional $1,500 donated by nurseryman and landscape designer Evert Asjes Jr. and the help of many individual donors, who volunteered both their time and money, WGC members along with some of the girls who resided at Crittenton completed a major portion of the landscaping of the center.[9]

As an aside, but an important one, Evert Asjes Jr. did much to help The Westport Garden Club, in addition to his generous donation to the

Crittenton Center. Back in 1951, Asjes had taken over the business of Rosehill Gardens, established by his father in 1914. The business had started with annuals, then expanded to trees, shrubs, and perennials of the highest quality. The senior Asjes's purpose was "to beautify Kansas City through gardening," a goal that meshed with that of The Westport Garden Club. He realized that in order to provide his customers with the best possible plants, he needed to grow them in local soil, a concept not previously adopted by nursery owners. Over the years, Evert Asjes Jr. contributed much to the WGC, loaning the women truckloads of beautiful plants for their flower festivals at the Nelson-Atkins, contributing to many club projects, and generously sharing his vast knowledge of horticulture and landscape design with WGC members.[10]

Figure 8.5 With the help of WGC members, Crittenton girls planted and raised a fruit and vegetable garden. The hortitherapy program was so helpful to the girls that the center decided to hire a hortitherapist and build a greenhouse. (WGCA)

After working with Evert Asjes Jr. and the teenagers at Crittenton, the WGC might have left the maintenance of the gardens up to the center. It was the club's practice to get a project rolling and let the organization follow up. However, after watching the Crittenton girls participate in the landscaping project and having recently learned about hortitherapy, Sallie Bet Watson, Marie Bell O'Hara, Sally Cowherd, and Norma Sutherland, along with some other WGC volunteers, decided to help the Crittenton residents plant a vegetable garden. A friendly farmer with a plow got the plot ready for the girls to seed and plant. According to Norma, at the beginning, many of the girls were silent and withdrawn, but within a few weeks they looked forward to working in the garden and considered participation

a privilege. Laura Kemper Carkener reported that as the WGC members "saw the young people's interest grow, our members moved indoors to create two rock gardens under lights. Before the summer was over, Crittenton residents were enjoying green beans and melons from the garden and zinnias cut from its border. And they were gathering for recreation around an indoor garden filled with tropical plants." Norma Sutherland wrote that working at Crittenton was "one of the most satisfying and exciting projects" on which the club had ever embarked.[11]

Norma Henry Sutherland: A Perfect Match

Seldom have the talents and interests of an individual aligned so perfectly with an organization's mission and purpose as did Norma Sutherland's with those of The Westport Garden Club and The Garden Club of America.

Figure 8.6 Norma Sutherland won many blue ribbons for her floral arrangements. (WGCA)

Documented in the Smithsonian Institution's Archives of American Gardens and featured in the GCA book *Gardens Private and Personal*, Norma's garden expressed her true love of gardening. She was a horticulture judge who grew trees and shrubs from cuttings as well as amaryllis and clivias from seed in her greenhouse. Norma is remembered to this day by club members for once exclaiming at a workshop, "I just love to propagate!"

In addition to serving as a floral design judge at many shows, Norma, a talented flower arranger in her own right, received numerous blue ribbons plus the GCA Certificate of Excellence in Flower Arranging.

She was honored to exhibit, along with Susie Vawter, at the Chelsea Flower Show in England, bringing home a bronze medal.

Garden photography was another of Norma's interests, her stunning work appearing on the cover of the GCA *Bulletin*.

Not surprisingly, other club members quickly recognized Norma's leadership skills and elected her club president. After serving as president, Norma went on to hold numerous jobs at the zone and national levels, culminating in a GCA vice presidency. She won many awards, including the Zone XI Creative Leadership Award and the Medal of Merit. At the 2005 GCA Annual Meeting, Norma served as an honorary chair.

Club members remember Norma for her boundless enthusiasm, wicked sense of humor, and willingness to teach and mentor others. A club award named in her honor recognizes novice members who demonstrate Norma's qualities.

Remaining as pretty and elegant in old age as she was when Gregory Peck chose her as the University of Kansas Yearbook Beauty Queen, Norma was a Westport Garden Club star.

In spring 1980, in honor of the observance of the traditional Arbor Day, GCA encouraged all of its 189 affiliated clubs to plant a tree somewhere in their respective communities. The WGC decided to do so at the Crittenton Center. In reporting on the event, Glenda-Jo Self, the garden editor of *The Kansas City Star*, included some sage advice offered by the WGC: "Don't plant a $25 tree in a $2 hole." In other words, after spending money for a tree, in order for it to prosper, it is worth taking the time to dig a hole that is big enough—one that is two to five times as wide as the root ball and no deeper than the root ball is tall. Club members had a chance to see the new tree when they held their October 1980 meeting at the Crittenton Center. They toured the facilities and gardens and listened to the girls themselves talk about the benefits of the horticulture program. While the appearance of the Crittenton Center improved with new landscaping and gardens, the real impact of the efforts affected the WGC volunteers and the Crittenton girls, all of whom found the work therapeutic. In addition to horticultural skills, planting and tending the gardens also taught values such as teamwork and perseverance. In January 1981, WGC Recording Secretary Virginia S.

Foresman reported that "one girl who had completed her stay at Crittenton just got a job at Rosehill Nursery!"[12]

In 1981, when Marie Bell O'Hara learned that Longview Community College was offering horticulture classes, the WGC underwrote the cost for ten girls from Crittenton to attend the six-week accredited program. WGC members also extended hortitherapy beyond the garden during the winter months, arriving at Crittenton on Wednesday afternoons to give girls instruction on planting narcissus, creating hanging baskets and terrariums, and making pine wreaths. In May 1981 Julia Tinsman taught a flower-arranging class to eleven girls, which was such a success that Marie Bell O'Hara presented another floral-arrangement demonstration later that same month. "Our mighty-mini project, the Crittenton Center, continues to flourish," President Barbara Seidlitz said in her 1981 annual report.[13]

The hortitherapy program at Crittenton proved to be so helpful to the girls that the center decided to hire a hortitherapist and build a Nature Science Center, for which The Westport Garden Club helped raise funds. In doing so, the club joined forces with the Kansas City chapter of Wellesley alumnae, who for years had held an annual garden tour to raise money for scholarships and their outreach program. WGC members offered to staff a boutique open during the evening before and the day of the tour. There they would sell plants, garden accessories, and collectibles—all to benefit Crittenton. As reported in *The Independent*, "The Westport Garden Club will add their panache to the 53rd annual Wellesley Garden Tour. The something new will be a Shoppe (offering choice things from antiques to et ceteras)." The garden tour, held on May 24, 1983, included a stop at the garden of WGC member Marie Bell O'Hara. WGC members coordinated sale items, raised plants, and staffed the shop—a major undertaking—but the boutique raised over $19,000. Crittenton used that money to help construct a greenhouse and a classroom specifically for horticulture therapy.[14]

With a science center, a dedicated hortitherapy program, and a hortitherapist on duty at Crittenton, The Westport Garden Club stepped back. Still, they held their September 1984 meeting at the center, and they continued to volunteer there occasionally and keep track of goings on at the center. In 2008, the WGC invited Andrea Salisbury, the director of the horticulture therapy program (HTP) at Crittenton, to refresh older club members' memory about the HTP and to acquaint new members with the special program that the WGC had started thirty years earlier. "The goal of the HTP," as Salisbury put it, "is to relieve stress, encourage problem solving, and achieve success, thus promoting self-esteem."[15]

Once set in motion, hortitherapy became an integral part of the Crittenton Center experience. In 2010, Krista Allen, the Development Officer of the Crittenton Children's Center, emailed WGC member Helen Lea to ask her to plan a time to introduce the center's new horticultural therapist to garden club members. The new hire had recently graduated from Kansas State with a BS in horticultural therapy and looked "forward to the opportunity to meet the ladies of The Westport Garden Club" responsible for the inception of the HTP.[16]

For Christmas 2015, instead of gaily wrapped jars of jams and jellies for retirement homes, the WGC decided to give garden gloves to the residents of the Crittenton Center.

Project 2000: The Children's Center Campus

In 1998, GCA asked all affiliate clubs to sponsor a Project 2000 to mark the start of the new millennium. For its project, the WGC chose to support the Children's Center Campus by building a special garden in the playground-courtyard used by children in the YWCA Nursery, the Nursery School for the Visually Impaired, and the Therapeutic Learning Center. Members then worked on two plans—one for the garden and one for the fundraising project necessary to ensure that they had the money to make the courtyard a beautiful, safe, and appealing place for children.

To raise funds, the membership voted in May 1998 to hold a House and Garden Tour featuring a garden boutique. Hazel Barton and Mariel Tyler Thompson headed the steering committee and set a tour date for May 5, 1999. Four club members graciously offered their homes and gardens for the tour—Betty Goodwin, Adele Hall, Sharon Orr, and Betty Thompson. Ellison "Kelly" Brent Lambert and her husband, Sandy, volunteered to set up the "Shop in the Garden" boutique in their garage and backyard, and offer a place to serve tea. The Lamberts immediately set to work painting and refurbishing their garage!

The WGC sent out invitations designed by artist Jack O'Hara, husband of club member Marie Bell O'Hara. The steering committee had determined that to make the event successful, club members, as well as the daughters

Figure 8.7 To raise funds for their Project 2000 at the Children's Center Campus, four WGC members, including Sharon Orr, offered their homes and gardens for a tour. (Kilroy)

and daughters-in-law they recruited to help, would all need to buy a ticket at the cost of forty dollars each. Members also were required to contribute an item for the Collector's Corner and donate a planted pot or basket to sell in the Shop in the Garden.

Thanks to all the thought and effort that went into it, the WGC's first House and Garden Tour was a resounding success. After two days of rain as WGC members set up for the garden tour, May 5th turned out to be a perfect day—sunny and bright. The allotted maximum of 540 tickets had sold out, and many of those who could not get a ticket made a contribution. In addition, the Shop in the Garden boutique sold almost everything in sight. The WGC now had sufficient funds on hand to set to work on the Children's Center Campus garden.

In lockstep with fundraising efforts, Betty Goodwin and her committee planned the new courtyard garden, but a children's garden, in a world of rules and regulations, presented several challenges. The state licensing board had to approve all plant materials; heavily scented flowers that might attract bees were not permitted, nor were poisonous or thorny plants. All the garden plantings and accessories needed to be clean, safe, and heat tolerant. Although an irrigation system was essential to maintain the plantings, its design required precision to avoid water on wheelchair ramps and sidewalks. Betty Goodwin reported on the project in November 1998: "We selected colorful plantings with textures and mild scents. Creating seasonal interest was important; ornamental trees, flowering shrubs, bulbs, herbs and annuals fill pocket gardens throughout the area. Vertical plantings on sturdy cedar trellises provide interest and shade. We have included spaces for children to dig, plant, and harvest."[17]

Along with new plantings and play structures, the club also wanted to convert one of the garden plots into a water feature, "adding the look, sound, and feel of moving water to the environment in a very child-safe manner." Wendy Powell, who had assumed command of the water feature, reported that plans for it, designed by Water's Edge in Lawrence, Kansas, had been submitted in 1998 to the Missouri Department of Health, Bureau of Child Care. The State of Missouri finally approved the addition of the water feature in October 2003, three years after the installation of most of the rest of the garden. Wendy noted that with the approval in place, a representative of the Children's Center Campus stated that before the club went ahead with

Figure 8.8 Betty Kennedy Goodwin and her committee planned the courtyard garden for the Children's Center. (WGCA)

the installation of the water feature, they still needed to determine how best to maintain it. "Let's remember," the ever-patient Wendy Powell ended a memo by saying, "Rome wasn't built in a day."[18]

Then, in January 2004, after waiting all that time for approval from the State of Missouri for a water element, the Children's Center board decided against it, saying they would rather have WGC members use their remaining funds (some $12,000) to improve existing beds and arbors, plant bulbs, and provide ongoing maintenance of the playground area. Duane Hoover of Kauffman Garden agreed to determine a maintenance plan to use the rest of the WGC funds, so that once these last plantings were finished, the WGC could turn the playground and its maintenance over to the Children's Center. "This will complete the project—originally 'Project 2000,'" Wendy Powell reported to the WGC Children's Center Project Committee in 2004![19]

Tulips on Troost: An Initiative for Change

WGC members became acquainted with Durwin Rice, a Kansas City Art Institute instructor, when he gave a demonstration/workshop for them on February 5, 2006, entitled "The Art of Decoupage with Botanicals." When, in the spring of that year, Rice proposed a project to beautify Troost Avenue, WGC members were interested. For decades, Troost Avenue acted as a dividing line between Kansas City's black population on the east side and the white population on the west. Rice's hope was to change the status quo and move beyond the racism that Troost Avenue represented. His idea was to plant Dutch tulip bulbs to honor the man for whom the street was named: Dr. Benoist Troost, a Dutchman who moved to Kansas City with his wife in the early nineteenth century. As the city's first doctor, Troost tended early residents—doing so regardless of their color or social status. The thought that underlay what became known as the Tulips on Troost Project was that, in a sense, Dr. Troost's spirit could again serve the city, this time by helping to heal the wound represented by Troost Avenue.

Within six weeks after launching the tulip project with the goal of planting 10,000 tulips in fall 2006, the planters had surpassed the original goal. The all-volunteer group reset their sights for the next year to plant 50,000 bulbs along Troost Avenue in the fall for spring beauty in 2007 and also to plant annuals and perennials along the avenue for spring and summer color. The objective of adding some color and beauty to Troost Avenue had sparked the imagination of the community. A very long list of institutions, businesses, and individuals donated money to purchase supplies and volunteered time to help perform the labor. Many neighbors along Troost offered their help in planting and maintaining the gardens.

The Westport Garden Club enthusiastically joined the list of participants in Tulips on Troost, with Carolyn Steele Kroh spearheading the club's effort. While the club donated $500, many club members volunteered to bring shovels and plant bulbs, including club President Sally West, Eulalie "Eulie" Zimmer, Ellen Jurden Hockaday, Lyndon Chamberlain, Jill Bunting, Margaret Hall, Nancy Stark, Blair Hyde, Dody Gates, Wendy Powell, Lorelei Gibson, Paget Higgins, Ginny McCanse, and Carolyn Kroh. As Durwin Rice stated, Tulips on Troost gave our city's residents "alternative, new and significant ways to view color" and provided "a

powerful symbol of hope, achievement, and healing." In 2010, with the recommendation of the WGC, Rice and the Tulips on Troost Project received a GCA Club Civic Improvement Commendation.[20]

The Joplin Project Blossoms After Tornado's Devastation

Joplin, Missouri, with a population of approximately 50,000 people, is located about two and half hours south of Kansas City, in the southwest Missouri Ozarks. The town, situated in an interesting topographical area featuring the Grand Falls and the federally protected Wildcat Glade, claims fame for native sons Langston Hughes and George Washington Carver. (The George Washington Carver National Monument is located there.) Joplin is also on historic Route 66 and is known for hiding Bonnie and Clyde from law enforcement in 1933. However, what riveted the nation's attention on Joplin on May 22, 2011, was a catastrophic EF5-rated multiple-vortex tornado, which killed 153 people and destroyed

Figure 8.9 The Tulips on Troost Project started with the goal of planting 10,000 tulips. (Kilroy)

everything in its path, including an estimated 400 commercial buildings and the city's largest medical center (and largest employer).[21]

WGC members voiced their concern immediately and just days after the tornado struck began pondering ways in which they might help the residents of this Missouri town. Looking at photos of the devastation, Jill Bunting expressed her dismay "that all the trees are simply gone" and suggested that she and her fellow club members might use "our ability to draw upon top arboretum expertise, wisely replanting trees for the future of Joplin." President Betty Kessinger suggested waiting until the initial shock and grief had subsided and then asking Joplin leaders "how we can help from a garden club approach."[22]

On September 12, 2011, at the first WGC meeting held since the tornado had struck, the club allocated $1,000 toward the Joplin Project. Barbara "Barb" Farmer Thompson and Eulie Zimmer volunteered to co-chair the endeavor. Barb, a Joplin native, contacted Chris Cotten, director of the Joplin Parks and Recreation Department, who suggested that replanting an entrance garden for Cunningham Park would boost the city's morale. Barb and Eulie headed to Joplin in November along with some other interested WGC members to start formulating a landscape plan for the thirty-nine-by eighteen-foot area. They also took potted mums to distribute to those displaced families still living in FEMA trailers. The following March, a contingent of club members went back to Joplin to evaluate and tweak the landscape plan they had drawn up, and in May, WGC volunteers, along with members of the Petal Pushers, a Joplin garden club, assembled to plant pink flowering crab trees, boxwood, golden flame spirea, and assorted annuals and perennials. The six crab trees planted by the gardening women were among the 153 trees planted in Joplin to commemorate the 153 individuals who had lost their lives in the tornado. On May 22, 2012, the one-year anniversary of the devastating natural disaster, the Cunningham Park entrance looked beautiful. Barb Thompson suggested to her fellow club members that they all go back to Joplin in the fall to plant irises, Joplin's city flower.[23]

However, the summer following the installation of the Cunningham Park garden turned out to be a season of terrible drought for Joplin and other parts of Missouri. Many of the plantings done by The Westport Garden Club did not survive. Undaunted, the club decided to try again.

Figure 8.10 The second plan for the Joplin garden called for drought-resistant native plants, including purple milkweed (*Asclepias purpurascens*), which attracts fritillary butterflies. (McCanse)

Diana Jackson Shand Kline led the charge: "I have begun the initial stages of restarting the Joplin Project," she reported to President Betty Kessinger in February 2013. Along with Jill Bunting, Susan Pierson, Laura Powell, Laura Sutherland, Wendy Powell, Barb Thompson, and Eulie Zimmer, Diana headed to Joplin to assess what plants they might salvage. Then they appealed to Alan Branhagen and Eric Tschanz of Powell Gardens to design a garden of drought-tolerant plant species. The plan included the use of natives: silver blade Missouri primrose, purple poppy mallow, bright edge yucca, butterfly weed, purple milkweed, pale purple coneflowers, aromatic asters, and prairie dropseed. WGC volunteers planned to travel to Joplin on May 20, 2013, to replant the garden in time for a memorial ceremony in the park scheduled for May 22 in commemoration of the second anniversary of the deadly tornado. Unfortunately, stormy weather prevented the trip, but the Joplin Parks and Recreation Department and local volunteers got the garden planted and vowed to maintain it. After making a follow-up trip, Diana Kline happily reported to her fellow club members that she had found the garden looking just the way Eric Tschanz, Alan Branhagen, and the WGC had planned it.[24]

It Takes Only One Strong Woman to Make a Difference

At the September 12, 2011, meeting of the WGC, Sharon Orr, chair of the WGC Garden History and Design Committee, reported that the Kansas City Museum had contacted her with a request to redesign and restore their existing gardens. The Kansas City Museum sits on a 3.5-acre historic property, the former home of lumber baron R. A. Long and his family. Completed in 1910, the original property included the mansion, called Corinthian Hall, a carriage house and paddock area, a horse trainer's home, a garden shed, a conservatory with a pergola, and a greenhouse. In 1940, the city converted the mansion into a public museum; in 1950, transformed the conservatory into a planetarium; and in the 1960s, removed the greenhouse.

By the time the museum contacted Sharon Orr, they had already planned for a major restoration and renovation of Corinthian Hall (which would not be completed until 2021) and the other outbuildings. They hoped the new gardens would initiate the renovation and reflect the museum's new mission statement: "The Kansas City Museum preserves, interprets, and celebrates Kansas City through collections, exhibitions, and bold programs that reflect the city's evolution and spirit, and engage visitors in unfolding stories about Kansas City's vibrant history, cultural heritage, and pride."[25]

Sharon intended to form a committee and ask for volunteers at the September 12, 2011, membership meeting, but at this same meeting Barb Thompson and Eulie Zimmer asked for volunteers and funds for the Joplin Project. Although the Kansas City Museum did not ask for monetary donations for the garden project, since its grounds and buildings were owned by the city, it did ask WGC members to volunteer their time and expertise. Sharon does not recollect whether she received any offers to help with this project, but she does remember having become completely absorbed in researching the history of the museum's gardens. "I kind of

Figure 8.11 Corinthian Hall at the Kansas City Museum was originally the home of lumber baron R. A. Long. (Orr)

fell in love with the place and did it on my own with the hopes that we would make it a project," she recalled. From old pictures and some still-existing bones, Sharon planned and then helped plant the front oval, two quatrefoils in the back, and the back gate area.[26]

Figure 8.12 Sharon Orr singlehandedly designed the new Georgian garden plantings that surround the museum. (Orr)

Figure 8.13 WGC members, including (from left) Susan Chadwick Pierson, Blair Hyde, Sharon Orr, Betty Askren Kessinger, and Gina Miller, attended the Garden Launch at the Kansas City Museum. (Photo courtesy of Sharon Orr)

At the garden club's annual meeting on June 6, 2012, President Betty Kessinger saluted Sharon for "her efforts to singlehandedly finish the research and design of the new Georgian garden plantings that surround the museum." Other members had attended the Garden Launch with Sharon at the Kansas City Museum on June 3, including Helen Lea, Susan Pierson, Blair Hyde, and Betty Goodwin. Executive Director Christopher Leitch thanked the club for its efforts: "All of us at the museum are so grateful for your interest in, and support of, our new garden initiative. Deep, abiding, and luscious thanks to the WGC for all they've done for us—especially you, Sharon Orr."[27]

Also at the year-end meeting, Betty related that she had sent in her annual president's report to GCA headquarters, including information about the Joplin and Kansas City Museum Projects. Even if only a small number of members had participated in each project, she noted, "where one of us

works in our name, we all work in our name. We are a club. We are The Westport Garden Club. We all believe in the same thing; we are all striving for the same ideal. We want to preserve and protect this land that God has provided us and His bountiful flowers and trees. We all declared a desire to share our love and knowledge of gardening; let's not forget what we are about."[28]

At the June 2013 meeting, Sharon showed photographs of how the gardens at the Kansas City Museum had appeared and what she and a few others had accomplished in the two years since the WGC had begun helping the museum with its garden renovation.

One Hundred Trees for One Hundred Years

To celebrate its one hundredth birthday in 2013, GCA launched the Centennial Tree Project, with the slogan "Preserving the Past, Growing the Future: Trees, Our Living Legacy." GCA encouraged each member club to propagate one hundred trees of one variety. The WGC's Centennial Tree

Celebration Committee, which included Wendy Powell, Dody Gates, and Lorelei Gibson, decided to propagate *Carya ovata*, shagbark hickory trees. Native to North America, shagbark hickories are common in the eastern United States, not so common in the central zone, but hardy in zones 4–8. (Kansas City is in zone 6.) Named for their interesting peeling bark, shagbark hickories are slow growers, yet easy to care for and can live for 150 years or more. Their long taproots make the trees drought-tolerant and reinforce soil, especially important along creek beds. The nuts of the shagbark hickory are edible, and the tree is beautiful both in the fall, when its leaves turn a brilliant yellow, and in the spring, when it blooms with yellowish-green flowers.

To help implement the Kansas City Centennial Tree Project, the WGC partnered with Powell Gardens, which provided a greenhouse and nuts from a parent tree, a magnificent shagbark hickory standing at the Gardens' entrance. Beginning in 2009, WGC members went to Powell Gardens for a series of propagation workshops. They stratified the nuts, planted, potted, and repotted the seedlings so that they would be ready to plant in 2013. Then, with help from homes associations in both Kansas and Missouri, and the Kansas City Parks and Recreation Department, WGC members identified planting sites for the one hundred trees throughout the Kansas City area, including along the creek in Peetwood Park in Mission Hills, Kansas.

The Loose Park Monarch and Shade Gardens

As we saw in the prior chapter, when Doug Ladd, the conservation director for The Nature Conservancy of Missouri, came to talk to the WGC in February 2014, he inspired those gathered to become active in promoting native plants and pollinators. Kathy Gates, with honorary member Bob Berkebile, took the lead in organizing a club task force. In spring 2014, the club formed the WGC Native Plant Initiative, which morphed into the Kansas City Native Plant Initiative (KCNPI), since renamed Deep Roots KC.[29]

Besides experimenting with native plants in their own gardens, club members decided to take an active role in promoting native plants by establishing a demonstration garden of pollinator plants in Loose Park across from the Garden Center. A National Fish and Wildlife Foundation (NFWF) grant partly funded the Loose Park Monarch Garden. The NFWF had started a Monarch Butterfly Conservation Fund in 2015 "to protect, conserve, and increase habitat for these iconic insects and other pollinators." In the twenty years prior to the founding of the monarch fund, the population of the North American monarch had plunged from one billion to fewer than sixty million, due primarily to a loss of critical habitat. NFWF felt it was crucial to become proactive to keep the monarch butterfly from becoming extinct. In its first year of existence, the Monarch Butterfly Conservation Fund received 115 proposals and total requests for more than $19.5 million to fund monarch projects. The NFWF granted money for 22 projects; the WGC's Monarch Garden was part of a package grant of $446,524 awarded to Kansas City Metropolitan Area Monarch Butterfly Conservation for the restoration of monarch habitat on 1,400 acres of both public and private land in Kansas and Missouri. Spearheaded by Burroughs Audubon Society of Greater Kansas City and supported by The Westport Garden Club, Kansas City Parks and Recreation, Grow Native!, Powell Gardens, and Deep Roots KC, the WGC project met the three key strategy objectives posed by the Monarch Butterfly Conservation Fund: "1) habitat restoration, 2) coordination and capacity building, and 3) seed supply and availability." Also of key importance, the Loose Park Monarch Garden is in an area open and free to the public and thereby serves as a model for stimulating native plantings and conservation.[30]

In spring 2016, The Westport Garden Club oversaw the planting of its demonstration garden, a full-sun native plant garden created to provide a food source for monarch butterflies on their migration. Alan Branhagen of Powell Gardens designed the 600-square-foot space populated by native plants that attract pollinators, including five species of milkweeds, which serve as host and nectar plants for the monarch: Spider milkweed, also called green milkweed (*Asclepias viridis*), is the first to bloom in the spring, with green and purple flowers, followed in June by rosy-colored purple milkweed (*Asclepias purpurascens*) and vibrant orange butterfly milkweed (*Asclepias tuberosa*). In July pink flowering Sullivan's milkweed (*Asclepias sullivantii*) blooms, and pink and white swamp milkweed (*Asclepias incarnata*) has blossoms in August and September. Besides the milkweeds, architectural plants in the Monarch Garden include wild bergamot (*Monarda fistulosa*), blue wild indigo (*Baptisia australis var. minor*), rattlesnake master (*Eryngium yuccifolium*), and curlytop ironweed (*Vernonia arkansana*). In the

center of the garden a trio of compass plants (*Silphium laciniatum*) produce flower stalks of up to eight feet tall, once established. Clumping grasses, various ephemera, seasonal plants, and groundcovers make up the rest of the well-planned native sun garden, which does, indeed, attract butterflies.

Figure 8.14 In spring 2016, the WGC oversaw the planting of the full-sun native plant garden. (WGCA)

Alan Branhagen and Kathy Gates supervised the installation, and WGC member Rosalyn "Rozzie" Hargis Motter assumed leadership of the maintenance crew—of utmost importance to the garden's success. In the fall of 2017, the Monarch Garden, so beautifully tended by WGC members, hosted two important events. On September 22, the WGC and the KCNPI hosted a celebration party to announce the contribution that WGC had made to KCNPI with funds from the very successful second "Entertaining Gardens" fundraiser, which had been held that May. On September 23, KCNPI partners presented the Monarch Migration Celebration. The sidewalk in Loose Park, from the colonnade on the east side of the park to the Laura Conyers Smith Municipal Rose Garden on the west, represented the monarchs' migratory pathway from Minneapolis to Mexico. Signs with monarch facts lined the route, and actual monarch caterpillars, chrysalises, and butterflies were on display with one chrysalis actually eclosing, to the delight of those watching.

Two years later, a generous gift from the Jean Mary Love Blackman family enabled the WGC to plant two additional native plant gardens, each approximately 250 square feet in size. Planted adjacent to the entrance of

Figure 8.15 *Left:* A grant from the National Fish and Wildlife Foundation partly funded the Loose Park Monarch Garden to attract more monarch butterflies to the area and help them along their way as they migrate. This monarch chose Carolyn Steele Kroh's garden. (Kilroy)

Figure 8.16 Cindy Rapelye Cowherd, Kathy Gates, and Laura Sutherland greeted those who attended the Monarch Migration Celebration at Loose Park in fall 2017. (WGCA)

the Loose Park Garden Center in a more shaded area, these gardens are almost entirely populated by shade-loving native plants.[31]

Since installation, The Westport Garden Club has maintained all three Loose Park gardens. Working in two teams on alternate Tuesdays from April through October, club volunteers pull weeds, stake taller plants such as the compass plant, and, as winter approaches, cut short some foliage, leaving grasses and seed heads for wildlife benefit. Although most WGC volunteers are familiar with many native plants, they are most likely to identify the species of those natives they use in their own gardens, such as asters, salvia, coneflowers, liatris, and penstemon, though not necessarily all the grasses, milkweeds, and wild ephemerals planted in the Loose Park gardens. Luckily, honorary WGC member Eric Tschanz, formerly of Powell Gardens, is almost always on hand to help identify which plants to weed out. One Tuesday, Ann Milton was pulling what she thought were weedy grasses when Eric stopped her and let her know that she was ripping out a desirable bottlebrush grass (*Elymus hystrix*). A short while later, in another area, Eric told Ann to pull something that looked very similar to what she had mistaken earlier for a weed. Eric acknowledged that she was correct; it also was bottlebrush grass. However, the grass in question this time was a "weed" in the sense that it was not growing where it should have been.

In 2021, Grow Native!, the marketing and educational program sector of the Missouri Prairie Foundation, selected the Loose Park monarch and shade gardens as Native Gardens of Excellence. The foundation, a private, nonprofit 501(c)(3) prairie-conservation organization and land trust founded in 1966, is governed by a volunteer board of directors who hail from all over Missouri—Jefferson City, Kansas City, Lee's Summit, Neosho, Peculiar, Springfield, St. Louis, Troy, Walnut Grove, Warrensburg—and who are concerned about the rapid decline of the prairie. Grow Native! issued a statement about why they chose the Loose Park gardens as excellent:

> The presence of a well-maintained, attractive native plant garden in one of the most visible areas of prestigious Loose Park is a mark of excellence in itself. The ongoing, dedicated maintenance over the last five years—in all types of weather—by the volunteers of The Westport Garden Club has made all the difference to the success of these three gardens.[32]

In 2022, WGC members completed work on a map and signage that identifies the native plants in all three gardens. A QR code allows access to an interactive website specific to the Monarch Garden and shade gardens.

While maintaining the Loose Park gardens and promoting native plants, WGC members have continued their concern for the monarch butterfly. In summer 2022, the International Union for Conservation of Nature (IUCN) listed the monarch as endangered. The western monarch population had declined by a devastating 99 percent since the early 1980s, and the eastern population by 85 percent. Both declines have resulted from habitat loss, climate change, and the widespread use of pesticides and herbicides. Kansas City is on the flyway of the eastern monarchs, which spend the summer in Canada and the northern part of the United States. After their stop in Kansas City, in late summer the indefatigable butterflies begin their 3,000-mile trip south to the pine forests in central Mexico, where they winter until spring, when they begin their journey back north.

Figure 8.17 In 2021, Grow Native! selected the WGC's Monarch Garden as a Native Garden of Excellence. (WGCA)

The milkweed plant, which supports the monarch caterpillars, is vital for the species' survival. Thus, besides encouraging the public to buy and plant milkweed, which is available at area nurseries and at the annual Deep Roots plant sales, Conservation Committee chairs Wendy Burcham and Laura Hammond urged WGC members to do their part to help the monarchs. At the September 2022 membership meeting, they handed out packets of milkweed seeds with planting instructions. Awareness and planting natives have had some effect on the monarch population. In 2023, the IUCN amended the Red List Assessment of the migratory monarch from endangered to vulnerable.

Olmsted 200 Project:
PARKS: Where Nature Meets Community

When the GCA announced a celebration of Frederick Law Olmsted's 200th birthday, which would occur in 2022, the national organization challenged all member clubs to find a way to honor this man whom many consider the "father of landscape architecture in the United States." GCA assigned its Garden History and Design Committee (GH&D) to oversee the Olmsted 200 Project, called PARKS: Where Nature Meets Community, and invited all 199 GCA clubs to connect the needs of their communities with Olmsted's vision and legacy of parks and shared open spaces.

Two members of the local club's GH&D Committee, Susan Small Spaulding and Wendy Powell, suggested a plan for taking club members on a series of hikes in a few of Kansas City's many preserves and parks with designated trails. They chose parks in different areas of the city that they thought would be of special interest to club members. In November 2021, on the coldest day of the season, about twenty hardy souls embarked on the first Olmsted hike in the Parkville Nature Sanctuary. The following April, members hiked in Minor Park, along the Blue River, a wonderful follow-up to a recent meeting program about conservation efforts on the river. The third hike, held on Earth Day, April 22, 2022, took members out on the Rocky Point Glades Trail, high on the ridge overlooking Swope Park. The guide provided a botanical tour of the many early-season wildflowers in bloom. While gathering knowledge and inspiration, hikers also gathered more than they had bargained for—seed ticks! These hikes "were meant to celebrate Olmsted's love of nature and the importance of open, green spaces and public parks," as Wendy Powell stated. "Both Susan and I felt the hikes were well received although the ticks were not!" The Olmsted hikes opened many members' eyes to places close to home where they had never been and provided great camaraderie for club members.[33]

WGC members enhanced their knowledge about the legacy of Frederick Law Olmsted by reading Justin Martin's *Genius of Place: The*

Figure 8.18 WGC volunteers working at the Monarch Garden took a break to celebrate Frederick Law Olmsted's 200th birthday with "Hats Off to Olmsted," a tribute to the man considered the father of landscape architecture in the United States. (WGCA)

Life of Frederick Law Olmsted, Abolitionist, Conservationist, and Designer of Central Park and participating in a book discussion thereof. Dede Petri, former GCA president, presented a program, via Zoom, about Olmsted's life, work, and times.

The Azalea Garden Celebrates
Seventy-five Years of Caring for the Landscape

As you may already have noticed, the WGC tends to overdo, and some would consider that might well have applied in the case of the Olmsted 200 Project. Besides devoting a program to Olmsted, reading his biography, and taking Olmsted hikes, the club formed a committee to investigate the renovation of an azalea garden in Loose Park as one of its major Olmsted projects. The azaleas, which had been transplanted in 1965 from the azalea collection of Harry V. Severs, had been neglected for over fifty years. Although overgrown and weed-ridden, some of the hardy azaleas still managed to bloom every spring.

The Olmsted Committee, headed by Carolyn Kroh, called on Eric Tschanz to offer his opinion on renovating the garden; he thought it would be a good project for the WGC, especially because of its proximity to the club's monarch and shade gardens. However, he suggested consulting Alan Branhagen, who after twenty years at Powell Gardens had moved to Minnesota. Branhagen's response was, "I know exactly what you are talking about—that well-established evergreen azalea grove at Loose Park is pretty wonderful—thriving through drought and wet years. It certainly is a strong framework for a renovated garden with additional plantings." Branhagen came to Kansas City, met with the committee, and recommended extending the beds and adding one hundred new azalea plants. However, the COVID-19 pandemic caused members to temporarily put the project on hold.[34]

In 2023, Carolyn Kroh resumed the investigation to determine what needed to be done to restore the azalea garden and how much it would cost. First, Carolyn determined that the Kansas City Parks and Recreation Department needed to repair the watering system in Loose Park, which would affect not only the azalea garden and the WGC's other gardens but also the gardens of the Hosta Society and the Asia Society. Then she called on the Boy Scouts, who helped clean up the garden as an Eagle Scout Project. In May 2023, The Westport Garden Club voted to allocate the funds necessary to renovate the azalea garden as its seventy-fifth anniversary project, provided that the watering system was in place first. After planting and other repair work scheduled for spring 2025, the Civic Improvement

Committee would then be in charge of maintaining all four of these gardens in Loose Park, certainly a testament to the WGC's commitment to green spaces and to Olmsted's vision.

Decorating Wreaths for Fun and Friends

Even though the WGC membership increased twice, from fifty to fifty-five and then to sixty members in 2010, the club had previously determined that flower festivals/extravaganzas were no longer a viable option for a club its size. WGC members agreed that they needed to find less elaborate ways to raise money. For several years, they had gathered to assemble Christmas wreaths, and in 2004 the wreath workshop evolved into an annual fundraiser. At the time the motion passed, each WGC member had to sell, decorate, and deliver three wreaths at $40 each or five swags at $20 each. Alternatively, a member could donate $100 to the Wreath Project. In 2006, the wreath sale raised $5,000, which the club donated to Kansas City Community Gardens (KCCG). Although the Wreath Project was successful, members objected to having to sell wreaths, sometimes in competition with the Boy Scouts, which did not seem right. In 2008, the club decided that its members should be required to purchase only one wreath or donate an equal amount. That year the wreath-making work took place in Ginger Owen's garage, and she offered lunch in her home. Marty Ross, a freelance garden writer for *The Kansas City Star*, wrote a wonderful story about the festive occasion. Even with only a one-wreath obligation, the project was still a successful money-maker.[35]

The Wreath Project developed into a series of events, with workshops in October and November to make bows for the wreaths and then a gathering after Thanksgiving to decorate the wreaths. In 2015, Bob Trapp of Trapp and Company offered to preorder fresh wreaths and provide space in his floral shop for their assemblage, probably to Ginger Owen's great relief. Creative expression became obvious, as members and their guests gathered to chat, decorate, and enjoy a casual lunch of Southwestern soup, corn bread, and drinks, which quickly became a tradition. In 2022, Co-Chairs Martha Lally Platt and Elizabeth "Beth" Ritchie Alm orchestrated

Figure 8.19 *Left:* The WGC's seventy-fifth anniversary project will restore and replant an azalea garden at Loose Park that was originally established in 1965. (Kilroy)

Figure 8.20 Kristie Wolferman's beautiful wreath is one of many decorated by members at the annual WGC workshop. (Wolferman)

a festive afternoon. Many members contributed to the buffet lunch, including Laura Hammond's husband, Charlie, who shopped for dogwood flower–shaped muffin tins in order to make special WGC corn muffins. (Yes, the men's auxiliary is still alive and well!)

The sale of the wreaths and a percentage of sales from Bob Trapp's shop allowed WGC members' funds to again benefit one of its partners. Always a friend to the WGC, Bob has done presentations, hosted workshops, assisted with the annual meeting in 2005, and made the wreath workshops very special occasions. The WGC proposed Trapp and Company for a GCA Club Civic Improvement Commendation, which he received in 2019.

Shopping, Tours, and Speakers Have Wide Appeal

Another unlikely but easy way that WGC members have raised money has been through purchases made at J. McLaughlin, a clothing store chain founded by brothers Jay and Kevin McLaughlin in 1977. J. McLaughlin supports local organizations across the country, including several GCA clubs. In 2017, the store on the Country Club Plaza in Kansas City, Missouri, began working with The Westport Garden Club, setting aside a percentage of purchases made during selected weekends to donate directly to the club's partners. What garden club woman does not like new clothes? A chair and co-chair oversee J. McLaughlin events, staged twice a year. In recent years the enthusiastic chairs, Sharon Orr followed by Kristin Colt Goodwin, previewed these events with a fashion show at the WGC meeting preceding the sale, with members modeling the store's latest arrivals. With the advent of the J. McLaughlin sales, the store's donations from it and the receipts from the Wreath Project are given to KCCG, Powell Gardens, and Deep Roots in an annual rotation.

Figure 8.21 Mary Harrison, Shelley Allen Preston, Allison Langstaff Harding, and Kristin Spicer Knight Patterson showed off the latest fall fashions from J. McLaughlin during a 2024 membership meeting at the Overland Park Arboretum and Botanical Gardens. (Peggy Kline Rooney)

In 2009, the WGC hosted a "Friends and Flowers" boutique and luncheon with featured speaker Ron Morgan, an internationally recognized floral designer. This format proved to be a highly successful fundraiser, so the WGC decided to alternate the speaker/luncheon/boutique format with raising funds through a garden tour. Thus, during the summer of 2011, the Executive Committee, in searching for gardens for the next fundraiser, toured thirty-six members' gardens. They determined that ten of them would be accessible; the eleventh would work only if golf carts could be used to escort visitors up the very steep driveway to the house of Sharon and Rich Orr! Eventually, Ginny McCanse and Marilyn Hebenstreit, co-chairs of the proposed event, called "Entertaining Gardens," chose six gardens—those of Lorelei Gibson, Margaret Hall, Betty Goodwin, Norma Sutherland, and Ginny's and Marilyn's own gardens. Each hostess would have a team of members working with her to present her garden in the best possible way and to ensure that it was an "entertaining garden," one properly set up for a garden party. Besides selling tickets, a shop at the Hebenstreits' house sold garden accessories and potted succulents, terrariums, and fairy gardens—all planted by WGC members.

After two years of planning, "Entertaining Gardens" took place on May 18, 2013. The honorary chairs of the fundraiser were to be WGC member Adele Hall and her husband, Don Hall, of Hallmark Cards. Adele Hall's unexpected death (this before the invitations had been sent) caused great sadness among club members, but they determined, and her family agreed, that she would have wanted her name to remain on the letterhead. The club chose Powell Gardens' Good to Grow program as the beneficiary of the event. "Designed to expand the impact of the Heartland Harvest Garden," the Good to Grow program would help children learn about nutrition and link growing food to food they eat. More than five thousand children each year could participate in planting seeds, watering and caring for plants, and harvesting fruits and vegetables, and then eating them. As the beneficiary of the fundraiser, Powell Gardens helped out with "Entertaining Gardens" by sending a horticulturist to each of the six houses to answer any gardening questions.[36]

Four years later, on May 13, 2017, the WGC again held an "Entertaining Gardens" event. This time, ticket holders could view four gardens, each with its own theme. The garden of Catherine "Kay" Strick Newell was

"An Homage to Italy"; Carolyn Kroh's, "An Artful Garden"; Lyndon Chamberlain's, "Cape Cod Charm"; and Sharon Orr's, "A Romantic Hilltop Retreat." (By the way, golf carts *were* made available to transport visitors up the Orrs' steep driveway.) A boutique at Ginger Owen's home featured unique garden accessories, tools, and plants. The club sold over five hundred tickets and collected $17,000 from underwriters. The tour was such an incredible success that this time around The Westport Garden Club was able to write a check for $45,000 to Deep Roots KC, the organization that WGC had started as the WGC Native Plant Initiative. These proceeds allowed Deep Roots KC to hire a full-time director to further the cause of promoting native plants in the Midwest.[37]

The next "Entertaining Gardens" fundraiser, slated for 2021, became a Zoom event due to the COVID-19 epidemic. In the virtual gathering, featured speaker Danielle Rollins presented entertaining and designing tips straight from her book *A Home for All Seasons: Gracious Living and Stylish Entertaining*, published in 2020.

Although the WGC had embraced the native plant movement and the environmental concerns of the 1960s and 1970s, gardens had not gone out of style. Garden tours and garden-design workshops still attract large numbers of people—not just gardeners but those interested in beauty, in green spaces, and in the peacefulness that comes from spending time in a garden.

Figure 8.22 *Left*: The "Entertaining Gardens" Committee chose Sharon Orr's garden to be among four gardens on the 2017 tour and used golf carts to transport visitors up the Orrs' very steep driveway. (Kilroy)

CHAPTER 9

It's All About Gardens

THE GCA MISSION STATEMENT BEGINS "The purpose of The Garden Club of America is to stimulate the knowledge and love of gardening," something the women of the WGC take seriously. However, as all gardeners know, loving gardens is far easier than loving gardening. Gardens demand a good deal of work, and true gardeners need to be knowledgeable about horticulture, growing conditions, and maintenance requirements. Gardens are also transitory and can be affected by many factors such as neglect, weather conditions (including periods of drought or damaging storms), and, of course, ravishing by deer, squirrels, rabbits, insects, and disease. Certainly, garden club members enjoy visiting gardens and sharing their own gardens with others, but they understand the effort involved not only in maintaining a garden but also in preserving gardens of historical significance.[1]

A unique way to promote interest in gardening came from WGC garden enthusiast Norma Sutherland. In the May 1985 GCA *Bulletin*, Norma wrote a short piece titled "Flower of the Week." She related that on Sunday evenings in seasons during which flowers in her garden were blooming, her husband, Dwight, had developed the habit of going out to the garden to select a flower to pick. Norma would then condition the flower by stripping foliage from its stem and placing the flower in cold water overnight in order to fully hydrate it and keep it fresh and vibrant. After putting the flower in a vase, she would record the common and botanical name of the species on a note card. Every Monday morning, Dwight would march into work with a

new nosegay, a practice that his office grew to expect and anticipate eagerly. "What a simple way to promote interest and the love of gardening," Norma Sutherland concluded in her article.[2]

Garden Tours as Social Events

For many years, a tour of other members' gardens provided an annual highlight for the WGC membership, and many of the club's social events revolved around garden tours. In June 1962, the occasion of Father's Day sparked a garden tour and dinner. Marie Bell O'Hara organized a tour of six gardens in 1974, including her own as well as those of Lillian Diveley, Millie Hoover, Sally Wood, Hattie Byers, and the country home of Norma Sutherland, where lunch was served. In 1977, members toured the gardens of Leila Cowherd, Julia Tinsman, Betty Thompson, and Helen Nelson, who also hosted a cocktail party for members and their husbands. A garden walk in May 1992 included stops at the homes of Sally and Tom Wood, Sallie Bet and Ray Watson, Laura and Craig Sutherland, and Barbara and Pete Seidlitz, who hosted an elegant cocktail party before dinner at The Kansas City Country Club. However, as the club became more absorbed in its environmental projects and its partnerships, the sharing of each other's gardens happened less often.

"Wine in the Garden" Events Evolve into Pop-Up Garden Tours

In 2011, after attending the GCA Annual Meeting, WGC President Blair Hyde brought back the idea of spur-of-the-moment, informal garden tours,

Figure 9.1 *Left:* Carolyn Kroh's blue perennial garden. (Kilroy)

Figure 9.2 LL's Garden is the second one that Lorelei Gibson had documented by the Smithsonian Archives of American Gardens. (Gibson)

the idea that a garden club member could send out a last-minute notice of a pop-up garden tour whenever she found her garden looking especially inviting or had something exceptionally beautiful blooming. No preparation was needed on the part of the hostess other than guiding club members through her garden during the designated visiting hours and sharing ideas.[3]

Figure 9.3 Japanese anemones bloom plentifully in the fall in Kansas City and have been a feature of several pop-up garden tours. (Wolferman)

announced just a day or two in advance. Kimberly "Kim" Kline Aliber and Alison Ward agreed to chair a new committee, which would seek out members who would like to either hold a gardening techniques workshop or host a "Wine in the Garden" tour, wine being a further enticement to visit a member's garden. Margaret Hall volunteered to host one such event and remembers that about thirty attended. Some of the older members, who had not understood that it was an informal, come-as-you-are kind of happening, dressed for a formal cocktail party, high heels and all. After one of the formally clad ladies took a bad fall while looking at the garden, subsequent "In the Garden" tours took wine out of the title and clearly billed these garden visits as casual affairs with no food or drink promised.

The "Wine in the Garden" tours evolved into "pop-up" garden tours, the term and the concept derived from another GCA meeting—a zone meeting hosted by the Garden Guild of Winnetka, Illinois. WGC delegates Lorelei Gibson and Cindy Cowherd returned from the October 2017 meeting with

WGC members have been invited many times to visit the gardens of Lorelei Gibson, usually in September to see her outstanding Japanese anemones, and of Carolyn and George Kroh, to marvel at her blue perennial garden and his magnificent vegetable plot. Carolyn remembers once hosting a pop-up garden tour in early October, right before the COVID-19 epidemic,

Figure 9.4 *Right:* A pop-up garden tour at Carolyn Kroh's house allowed visitors to roam her one-acre plot, on which she has designed various garden rooms. (Carolyn Kroh)

when, for the first and only time, her Japanese anemones, monkshood, and cleome all bloomed at once. "It was glorious," she recalls, "and usually my garden is very uninteresting in late fall."[4]

Figure 9.5 Included in the pop-up garden tour at the Krohs' home was George Kroh's vegetable garden, for which he received a GCA Club Horticulture Commendation in 2021. (Kroh)

Although both the Gibson and Kroh gardens have been submitted to the Archives of American Gardens at the Smithsonian, having an archival garden or an estate garden is certainly not a prerequisite to hosting a pop-up garden tour. Visiting Helen Lea's garden is always a delight. Her home is filled with her paintings, and her small garden is expanded by dense plantings and the strategic placement of mirrors to fool the eye, a clever trick she learned after visiting gardens in England. In another example, Martha Platt's garden creates the effect of a comfortable "outdoor room" surrounded by layers of planting beds and walkways that she developed in keeping with the natural topography of the land. One year, when the glorious quince

hedge was blooming at the historic 1929 condominium building known as the Walnuts, several WGC members—Dody Gates Everist, Sally West, and Paget Higgins—who reside there hosted a tour of the ten acres of grounds and gardens. For her part, Margaret Hall claims to have hosted at least seven pop-up tours of the garden at her home, usually in the spring, when her tulips are at their peak, but also on any given day when she goes out into her garden and thinks, "It is so pretty and photogenic, I'd love to share this with friends." In recent years, pop-ups have replaced formal garden tours except for fundraising events.[5]

Figure 9.6 Helen Lea's small garden is densely planted, with mirrors to augment its beauty. (Kilroy)

Focus on Native Plants

Since the talk given to the club by Doug Ladd of The Nature Conservancy in 2014 and the creation later that year of the WGC Native Plant Initiative, now Deep Roots KC, the WGC's emphasis in gardening has focused not only on beauty but also on native plants and pollinators. Linda Evans and Susan Ambler Spencer, as 2017 chairs of the Conservation Committee, stated the committee's goal for the year was "to enhance and continue to support the health and future of pollinators in our region. It is so vital to educate and do the planting and tending of our gardens to help the bees and butterflies." The 2018 GCA Annual Meeting focused on the club's Healthy Yard Pledge, and the WGC Conservation Committee encouraged all members to take the pledge, vowing to care for their yards "without synthetic pesticides, weed killers and fertilizers except on rare occasions to resolve an infestation or to improve habitats for native plants and wildlife." Living up to the pledge proved to be another example of members of GCA thinking globally but acting locally.[6]

In November 2022, GCA launched its Native Plant Month Initiative, aimed at "creating public awareness of the importance of native plants." GCA clubs across the country asked governors of all fifty states to issue proclamations declaring April as Native Plant Month. The WGC participated in the initiative by requesting proclamations from governors of both Missouri and Kansas. As it turned out, Kansas already had designated September as its native plant month, with the first proclamation made by the governor of Kansas in 2007 and renewed each year thereafter. However, Missouri had not made such a proclamation, so the WGC, along with garden clubs in St. Louis, requested and gained the support of Governor Michael L. Parson. By April 2023, forty-eight states and the District of Columbia had declared April or another month as Native Plant Month. GCA announced: "We will continue these efforts in the coming year to build on the success of the initiative," always with the goal in mind of educating the public about the importance of native plants.[7]

At the 2023 GCA Annual Meeting, when honorary WGC member Bob Berkebile accepted the Cynthia Pratt Laughlin Medal, which is awarded "for outstanding achievement in environmental protection and the maintenance of the quality of life," he posed two questions to gardeners and garden clubs across the country: "What else can you or your club do to make an impact" in protecting the environment?, and "How do we define beauty in our gardens? If your garden does not sequester carbon, can it be considered beautiful?" Known for his advocacy for conservation in the urban landscape and soil regeneration, for helping to form the US Green Building Council, and as the founding chair of the American Institute of Architecture National Committee on the Environment, Bob purports that the idea of a beautiful garden has changed and expanded as more people have become aware that what they plant matters. Certainly, Deep Roots has made a difference, locally and perhaps even globally, as Bob Berkebile indicated in his acceptance remarks. In 2023, an event sponsored by Deep Roots, called "Cocktails for Conservation," took place in the native plant garden of WGC member Wendy Hockaday Burcham and her husband, Grant, and the event was repeated in another native plant garden in 2024.[8]

A Glimpse of Gardens Elsewhere

GCA members also have the opportunity to visit and tour gardens beyond their locales, thanks to the Visiting Gardens Committee and video tours on the GCA website. The GCA Visiting Gardens Committee plans trips both in the United States and abroad in order "to educate members of GCA clubs in garden history and design, horticulture, and the environment." In 2015, for example, GCA Visiting Gardens trips included venues in Portland, Oregon; Cuba; and Tuscany. Members interested in these domestic and international trips must act quickly, however, as they generally sell out on the same day they are posted![9]

The GCA Visiting Gardens Committee also facilitates club member visits to the gardens of GCA members throughout the United States and to the gardens of GCA international courtesy clubs. WGC members had the opportunity to participate in two special Visiting Gardens trips in 2011: a three-day, two-night "Gardens of New York City" trip in 2012 and a visit to California in April 2014, organized around an invitation from the Pasadena Garden Club. In 2024, Paget Higgins, WGC's Visiting Gardens chair, planned a February trip to Mountain Lake, Florida, where two WGC members—Diana Kline and Laura Powell—have homes. On this visit, WGC members spent a day at Bok Tower Gardens, where the azaleas were in full bloom, toured several private gardens, and saw unusual birds and wildlife on a nature hike.

The WGC Visiting Gardens chair can also arrange for individual members to tour the gardens of GCA members in other cities. For example, in 2008 the WGC Visiting Gardens chair helped Helen Lea see GCA members' gardens when she spent time in Portland, Oregon. Likewise, Eulalie Zimmer's trip to Santa Fe included visits to GCA gardens.

Interstate 70 trips between Kansas City and St. Louis used to occur with some regularity. In October 1996, the St. Louis Garden Club invited WGC members to spend two days visiting and touring gardens in their area. During COVID-19, the bond between the two clubs was resumed through a Zoom horticultural book club meeting. In spring 2022, when some ladies from the Ladue Garden Club visited Kansas City, Paget Higgins arranged for them to see the gardens of Carolyn Kroh, Lorelei Gibson, and Margaret Hall and to visit Powell Gardens.

Figure 9.7 Nineteen WGC members toured Bok Tower Gardens in Florida in February 2024 (*from left*): Wendy Powell, Charlotte Russell White, Martha Platt, a garden guide, Dody Everist, Peggy Rooney, Susie Campbell Heddens, Diana Kline, Susan Small Spaulding, Laura Keller Powell, Mary Ann Powell, Susan Ambler Spencer, Paget Gates Higgins, Kristie Wolferman, the garden director, Lyndon Gustin Chamberlain, Kay Strick Newell, Nancy Lee Smith Kemper, Lorelei Gibson, Gina Miller, and Pam Gyllenborg. (Photo courtesy of Peggy Rooney)

Garden History and Design

One of GCA's purposes is to maintain, restore, and document historic gardens. In 1914, just a year after its founding, GCA established the Historic Gardens Committee. In 1919 a Slide Committee began collecting files and glass lantern slides of historic gardens dating back to colonial times. To celebrate its seventy-fifth anniversary in 1988, GCA decided to donate its slide collection of 3,000 hand-colored glass lantern slides from the 1920s and 1930s and 35,000 35-millimeter slides to the Smithsonian Institution. The GCA collection became the nucleus of the Smithsonian's Archives of American Gardens (AAG), which now includes 65,000 photographic images and records that document 6,500 historic and contemporary gardens throughout the United States. These records are available to gardeners, landscape designers, and historians at the AAG office in Washington, DC, where they are being digitized and made available to the general public. The official gift deed from GCA to the Smithsonian, signed in 1992, also marked the year of the founding of the GCA Garden History and Design (GH&D) Committee.[10]

The national GH&D Committee asked each member club to designate a GH&D committee and chair. The purpose of the local-club committee would be to seek out photographs and information on historic gardens in their respective areas as well as to photograph and document contemporary gardens. Once again, The Westport Garden Club rose to the occasion, and its GH&D Committee members "discovered" a treasure trove of color plate slides of historic gardens in the Special Collections of the Kansas City, Missouri, Public Library. The women had come across the work of Frank Lauder. After losing his job during the Great Depression, Lauder took up photography, and in 1932 and 1933 he took more than 1,200 photographs of Kansas City homes, gardens, and landmarks. He used autochrome plates, a process patented by the Lumiere brothers in France in 1903 and the only way to make color photos until the invention of Kodachrome film in 1935. WGC member Norma Sutherland, through her family's Sutherland Foundation, gave the Kansas City Public Library money to scan Lauder's slides. The library's Stuart Hinds digitized them in 2000 so that they could be displayed on the library website and made available for the public to see as Autochromes: Frank Lauder Collection. The WGC submitted the slides themselves to the Smithsonian's AAG. That done, Norma initiated a project for the WGC to take present-day photographs of gardens featured in the old slides, this in the hope of providing a then-and-now perspective. However, the WGC GH&D Committee experienced limited success convincing homeowners to participate. Having received a letter of inquiry from

GH&D Chair Boots Leiter, the owner of the former M. B. Nelson house at 5500 Ward Parkway simply refused to reply. In 2001, Norma Sutherland received the GCA Club Historic Preservation Award in recognition of her work in submitting the Lauder slides to the Smithsonian and initiating the then-and-now project.[11]

GCA made clear that "preserving our American garden legacy" did not limit documentation to grand gardens designed by professionals but encouraged submission of all manner of gardens, including vegetable gardens, balcony gardens, and apartment and townhouse gardens. In 2009 Joyce Connolly, a museum specialist at AAG, reminded GH&D chairs,

> We are losing important historical information for future researchers if we focus on just a certain "type" of garden or ignore prevalent design trends. Not to make too light of the issue, but AAG doesn't have a single garden image with a pink flamingo in it! While many are thankful that this design trend has pretty much disappeared, it's important to consider that we're not arbiters of taste, but documenters of garden history.[12]

Figure 9.8 Not-entirely-pink flamingos add a touch of whimsy to Sharon Orr's garden. (Kilroy)

Ten Gardens Will Live On for Posterity

When Boots Leiter became the club's GH&D chair in 2004, she did not mention gardens with flamingos but did ask for ideas of gardens that could and should be documented to send in to the AAG. She explained the rigorous process involved and the need for consent and cooperation from the garden owners. While the WGC prepared for the 2005 Annual Meeting, Boots and her committee began selecting photos of two gardens to submit to AAG—the garden of WGC member Jan Dye and husband Ned Riss, called Gracie's Garden; and Teckel Hall, the garden of Cindy and Dwight Sutherland Jr. Committee member Jill Bunting wrote the description of Teckel Hall as the other committee members began working on documentation of a third garden, that of Norma and Dwight Sutherland.

In November 2005, Boots attended a GCA GH&D meeting in Lake Forest, Illinois. She hand-carried the Riss application for Gracie's Garden, wanting to certify that it met all the criteria; it would be the first submission from the WGC. On January 1, 2006, the WGC learned that the Smithsonian had accepted the Gracie's Garden submission, which was listed as AAG Garden #KS021, the twenty-first garden documented in the state of Kansas. Gracie's Garden, named for the Risses' dog, is on an estate at 2435 Drury Lane in Mission Hills, Kansas. The home originally belonged to J. H. Revelry, one of eleven prestigious Kansas Citians approached by developer J. C. Nichols to build in his new suburb, Mission Hills. In 1927, Revelry retained the architect Edward W. Tanner to design his large Tudor home. On the west side of the property an English rill meandered from a tea house to a small pond, adjacent to a tennis court, all of which needed restoration when the Risses purchased the property in 1975. Originally, the property had no gardens but perhaps the largest variety of trees on any Kansas City residential site—including hackberries, buckeyes, black walnuts, sycamores, and golden raintrees. The Risses added to the arboretum, planting sixty different varieties of trees, as well as a wealth of shrubbery, perennial beds, and an assortment of antique and David Austin roses. They added a water garden and lotus pond in 1998. Gracie's Garden was one of the private gardens toured by delegates to the 2005 GCA Annual Meeting in Kansas City.[13]

In 2006, the Smithsonian AAG collection admitted Teckel Hall, at 2411 West Fifty-Ninth Street, the garden of Cindy and Dwight Sutherland Jr.

Figure 9.9 The garden of Jan Dye Riss, called Gracie's Garden, was the first to be submitted to the Archives of American Gardens by the WGC. (WGCA)

Figure 9.10 Numerous gardens could be seen at Heron Haven, the home of Norma and Dwight Sutherland, which spanned sixty acres along Indian Creek in Overland Park, Kansas. (AAG)

Their house, designed by architect Edward Tanner in 1932 for Bruce Dodson Jr., an insurance pioneer, included grounds planned by the renowned landscape architects Hare & Hare. An article on the front page of the July 31, 1932, *Kansas City Star* features the layout of the Dodsons' English home on a one-acre lot. The garden plans highlighted and incorporated an unusually fine, spreading walnut tree and featured three levels of gardens with native stone steps down the hillside.

AAG also accepted the Norma and Dwight Sutherland garden, Heron Haven, at 10601 Nall Avenue in Overland Park, Kansas, in fall 2006. Norma, Jill Bunting, and Jan Riss completed the research and documentation of the Sutherlands' sixty acres along Indian Creek. Photos of the entrance gate covered with New Dawn roses and of the driveway lined with tree peonies, euphorbia, daffodils, tulips, irises, day lilies, and herbaceous peonies competed with the colorful cutting and herb garden, the rose garden, the vegetable garden, and the greenhouse.

When WGC member Betty Goodwin became national chair of GCA's Garden History and Design Committee in 2006, she continued to encourage

WGC members to submit applications for their gardens to become part of the Smithsonian Collection. Lorelei Gibson needed no encouragement; she had already begun taking photographs and documenting her garden. Called the Horn Garden for the original owners of the property, John and Phoebe Horn, Lorelei's garden had also been on the 2005 Annual Meeting tour. Lorelei had completed extensive research on the estate, which is listed on the National Register of Historic Places.

John Horn was one of the eleven prestigious men chosen by J. C. Nichols to own an estate in Mission Hills. With Edward Tanner as their architect, Horn and his wife built their home in 1930 on a seven-acre hillside site overlooking a creek and farm fields. The Horn Garden, designed by the Hare & Hare landscape architectural firm, fulfilled developer Nichols's vision of "formal to frontier" land features, with formal gardens around the house yielding to wooded hillsides in three directions, suggesting a natural wilderness. A manmade rill cascaded down to the creek below.

In the 1950s the Horns sold the Wenonga Road property to Herbert O. and Margot Peet, a founding member of The Westport Garden Club. After her husband's death in 1984, Margot tried to keep up the large estate but could not. After its initial years of glory, the Horn Garden had fallen into disrepair when WGC member Lorelei Gibson and her husband, David, bought the property in 1997. They had barely moved in, and Lorelei had only a vague plan in mind for revamping the garden, when WGC members urged her to hold a fundraiser on her new estate. She acquiesced with her typical grace and calm disposition.[14]

Figure 9.11 In 1930, John and Phoebe Horn built a home in Mission Hills, Kansas, on a seven-acre hillside. Their garden fulfilled developer J. C. Nichols's vision of "formal to frontier" and was the first garden in Kansas to be documented by the Smithsonian. (WGCA)

Figure 9.12 In the 1950s the Horns sold their property to Herbert O. and Margot Munger Peet, a founding member of the WGC. (WGCA)

Lorelei's first tasks in the remediation of the landscape included tending to the trees, the overgrown yews, and the invasive vines, honeysuckle, and poison ivy. Restoring the perennial gardens proved especially taxing, as soil and plants had to be hauled up or down sets of stone steps. Lorelei would spend eight years bringing the gardens back to their former glory, preserving and enhancing the original design and planting more appropriate varieties of trees and plants that could withstand harsh Midwestern weather conditions. The restored estate garden of Lorelei and David Gibson, located at 6624 Wenonga Road, became part of the GCA collection in the AAG in February 2008.[15]

Nancy Lee Kemper took over as WGC Garden History and Design chair in 2009 and identified several gardens as good prospects for submission to the AAG. Wendy Powell and Margaret Hall had already begun documenting the garden of Virginia and Sarah Weatherly. For more than half a century, from 1943 to 1995, the Weatherly sisters gardened on a quarter-acre plot in the heart of Kansas City, Missouri, at the corner of Main and Fifty-Seventh Street. Around their "shirtwaist" house (half native stone and half wood frame), the sisters planted a paradise of garden rooms with raised beds of vegetables, herbs, and perennials. A long stretch of grass on the west side of the house was bordered by shrubs, perennials, and annuals, with hybrid roses climbing up the west wall of the house. Cort Sinnes, editor of *Flower and Garden* magazine, wrote about the Weatherly sisters, whom he called the "Sister Women"; they became known locally, nationally, and

internationally. In an article about the Weatherlys' garden published in *The Kansas City Star*, George Gurley opined, "It may be only a postage stamp garden. But when love and knowledge meet, small things can become immense." The garden of Virginia and Sarah Weatherly became a part of the GCA collection at the Smithsonian in 2011.[16]

Figure 9.13 The Weatherly sisters planted a series of garden rooms with beds of vegetables, herbs, and perennials. (AAG)

Laura Babcock Sutherland worked with Stacey and Reed Dillon to document their garden in Lawrence, Kansas. Wendy Powell and Nancy Lee Kemper toured the Dillon garden with Laura in late spring 2010. They liked "the thoughtful structure, plantings, and entire design" and thought they would be ready to submit the garden documents by spring of 2011. However, it was not until September 11, 2014, that the AAG added the Dillon residence as Garden #KS034; it was the first garden from Lawrence, Kansas, documented in the GCA Collection at the Smithsonian. Laura Sutherland presented a program on the documentation of the Dillon garden at the May 2015 meeting of the WGC.[17]

The Westport Garden Club GH&D Committee members documented the Rainbow Garden in Mission Hills in 2000. Designed and planted by WGC member Betty Goodwin between 1996 and 1998, many of the plants and shrubs used in the garden were native to western Missouri and eastern Kansas. The garden plan focused on providing continual interest throughout spring, summer, and fall, when blue asters attracted hundreds of monarch butterflies. A unique feature, a dry creek bed along the western property line, provided a natural conduit for spring and fall rains and an ideal setting for bulbs and shrubs.

Figure 9.14 Laura Babcock Sutherland worked with Stacey and Reed Dillon to document their Lawrence, Kansas, garden·for the Smithsonian. (AAG)

Figure 9.15 Betty Goodwin designed her garden using native plants and shrubs. A dry creek bed along the western property line provided a natural conduit for spring rains. (AAG)

In 2015, the WGC Garden History and Design Committee—consisting of Dody Gates, Alison Ward, and Lyndon Chamberlain—sent in to AAG photographs and documentation for another garden designed by Lorelei Gibson. In 2010, when Lorelei and David moved from their seven-acre Wenonga address to their new home at 3505 West Sixty-Fourth Street in Mission Hills, Kansas, the house had no garden. In fact, whoever had designed the large brick house on an irregular-sized lot had given little thought to the use of the outdoor areas. Even though her new property was much smaller than her previous estate, Lorelei still wanted to have a variety of gardens and space for the couple's collection of garden sculptures and mill wheels. Therefore, she conceived the idea of creating multiple garden rooms, bisected by brick walkways, in their 3,000-square-foot side yard. The Gibsons received notification on November 30, 2015, that their garden, documented as "LL's Garden," had been accepted into the GCA Collection in the Smithsonian Institution's Archives of American Gardens. Lorelei Gibson is the only WGC member to have two of her gardens documented at the Smithsonian.[18]

The WGC GH&D committee also documented Marvin and Emelie Snyder's garden of 250 dwarf conifers. Marvin Snyder had an exceptional eye and knowledge of dwarf conifers. He had served on the board of the American Conifer Society and as its president for two terms. He gave a talk to WGC members in November 1996 titled "Conifers for Our Heartland Gardens" and arranged for members to tour his garden. Snyder was a registered architect and a mechanical engineer and had worked for Butler Manufacturing Company as an employee or consultant for sixty-six years. Yet, when he died in 2021 his obituary focused on his gardening skills: "Marvin Kinyon Snyder, an avid local gardener whose personal garden is included in the Smithsonian Archives of American Gardens, passed away on March 31, 2121," just a month after his garden was accepted into the Smithsonian archives.[19]

The GCA website in November 2021 featured the Fairway, Kansas, garden of Carolyn and George Kroh, which they had successfully submitted to the AAG in May. When the Krohs moved into their home in 1988, the front yard had knee-high weeds, according to Carolyn, but several handsome old trees. The Krohs transformed the front of their house by adding dogwoods under-planted with white azaleas and Maureen tulips, boxwood hedges, and ivy borders. The Krohs' one-acre plot is knit together with stone and brick paths and artistic garden accessories, including ceramic birdhouses made by grandchildren. While Carolyn designed rooms for her shaded woodland garden, east garden, English cottage–style perennial garden, and herb garden, George made his vegetable garden "a room of his own." His efforts led to the WGC nominating him for a GCA Club Horticulture Commendation, presented in 2021.

Figure 9.16 In 2021, the Krohs successfully submitted their garden to the AAG. Carolyn's shaded woodland garden was featured on the GCA website. (Kroh)

The WGC continues to seek area gardens to document in addition to the ten already accepted by the Smithsonian. The club's documentation of the Beanstalk Children's Garden at Kansas City Community Gardens, when completed, will be in the Community of Gardens Story Collection in the AAG.[20]

GH&D Committee Travels Through History

The Garden History and Design Committee also educates WGC members about the history of gardens. At club meetings in 2022–2023, GH&D Co-Chairs Susan Spaulding and Jo Missildine, and members Jill Bunting, Carolyn Kroh, Rozzie Motter, Wendy Powell, and Sally West presented a series of reports on iconic gardens in the Kansas City area. In September, Eric Tschanz reported on the Ewing and Muriel Kauffman Memorial Garden on Rockhill Road, a garden conceived by Julia Irene Kauffman to honor her parents. Working in partnership with the Kansas City Parks and Recreation Department and Powell Gardens, the Kauffman perennial garden opened in 2000. Tschanz led a tour of the garden in May 2023 after a WGC meeting.

Figure 9.17 Listed on the Smithsonian's website as an historic city landmark, Verona Columns Park in Mission Hills features eight twelve-foot-high marble columns and a Carrara marble waterfowl fountain that landscape architect S. Herbert Hare brought back from Italy at the request of developer J. C. Nichols. (Kilroy)

In November 2022, Sally West explained the history of the Pilgrim Labyrinth and Butterfly Garden in Hyde Park. Although the garden is fairly new, having opened in 2017, its history goes back to the platting of Hyde Park in 1886. George Kessler designed a walking garden around Pilgrim Chapel (now Pilgrim Center), founded as a chapel for the deaf in 1941. Residents of Hyde Park raised money to build the butterfly garden and labyrinth to provide a place for walking meditation and quiet surrounded by native flowers and plants.

In February 2023, Jo Missildine reported on the Verona Columns Park, Mission Hills's most distinctive iconic garden. The park, listed on the Smithsonian website as an historic city landmark, is located at the circle

Figure 9.18 *Left*: In 2010 Lorelei and David Gibson moved from the Horn/Peet estate to a home on Sixty-Fourth Street, where, on a much smaller property, Lorelei created multiple gardens. (Gibson)

intersection of Ensley Lane, Mission Drive, and Overhill Road. When J. C. Nichols was laying out Mission Hills, he commissioned Hare & Hare landscape architects to design this parklet as part of his Country Club District development and specifically asked S. Herbert Hare to travel to Europe in 1924 to search for art pieces for this site. In Rome, Hare found an antique Carrara marble waterfowl fountain; in Venice he came across eight Verona marble columns, each standing twelve feet high, as well as a large white marble vase. The development of Verona Columns Park took more than a year. During the process, developers diverted a creek, built flagstone paths, and constructed steps from native stone. Abundant planting of both evergreens and beds of perennials and annuals completed the plans. By the way, S. Herbert Hare had studied at Harvard under Professor Frederick Law Olmsted Jr.

Jeanne Forney Bleakley, Marguerite Munger Peet, Helen J. Jones Lea, and Laura Lee Carkener Grace: Artists in the Garden

Artistic expression takes many forms, and a love of gardening and flowers has provided inspiration for many WGC members. The club has counted garden designers, sculptors, floral designers, needle arts designers, textile artists, photographers, ceramicists, jewelry designers, and painters among its members. Each of the four painters highlighted here sought and found inspiration in the garden.

Jeanne Forney Bleakley

Jeanne Bleakley's legacy of colorful paintings and vibrant landscapes that grace many Kansas City homes reflects her sunny disposition and love of flowers. Jeanne was a native Kansas Citian, graduating from Southwest High School and earning an art scholarship to Gulf Park Junior College in Gulfport, Mississippi. Later she attended the University of Missouri, the Kansas City Art Institute, and Johnson County Community College. Jeanne designed cards for Hallmark Cards, Inc., and had frequent one-person shows throughout the metro area. Her association with the WGC gave her much pleasure, and she was always happy to share her artistic talents whenever the club needed them for a fundraiser or similar event.

Figure 9.19 Jeanne Forney Bleakley selflessly shared her artistic talents with the WGC. (Courtesy of David Bleakley)

Marguerite Munger Peet

Marguerite "Margot" Peet was many things: a philanthropist, a legendary hostess, a knowledgeable gardener, and a loyal friend. But the one-word epitaph she chose for her grave marker in the Pantheon at Kansas City's Forest Hill Cemetery was "Artist."

From childhood, Margot dreamed of being an artist. After completing high school at The Barstow School, she intended to study art at Smith College. However, when family finances prevented her from attending college, she went to New York City, where she studied with pastel artist Clinton Peters. At the time she was primarily doing portraits. Reluctantly returning to Kansas City, again because of finances, she met Herbert O. Peet, who was smitten by her talent, joie de vivre, and beauty. (A long-lost, but supposedly reputable, source named Margot as one of the twenty most beautiful women in America.) The couple married shortly thereafter, and Margot began art studies with Randall Davey, who instructed her in the techniques of oil painting as well as advanced color theory. Her most noted teacher, however, was none other than Thomas Hart Benton, from whom she learned sculptural form and the use of egg tempera. She absorbed Benton's tenets of strong design and working from life study. Landscapes, still lifes, and views of her garden rounded out the subject matter of Margot's later

work. She continued to paint at her homes in Kansas City and Jamaica until 1983, when she suffered a debilitating stroke.

The 450 paintings found in Margot's attic after her death in 1995 attest to her calling as an artist. Not even her family had been aware of the scope of her work.

Figure 9.20 Known as a philanthropist, legendary hostess, and founding WGC member, Margot Peet chose a one-word epitaph for her grave marker: "Artist." (Self-portrait by Margot Peet)

Helen J. Jones Lea

From painting at her easel in kindergarten at Border Star School to spending solitary mornings painting at Monet's gardens at Giverny, Helen Lea's journey as an artist has been international in scope and astounding in variety. Helen studied art history and photography at Smith College, followed by further study in classical painting at the University of Perugia in Italy.

Twenty years later Helen received her BFA from the Kansas City Art Institute, studying under Walter Niewald. She spent a year as an artist at Hallmark, but the lure of a career in New York was hard to resist. She was a stewardess for legendary Pan Am Airlines; a New

York model; a staff member at the Frick Collection; and a designer of textiles for Martex, Missoni, and Fieldcrest, as well as paper goods for C.R. Gibson Creative Papers and puzzles for Ravensburger.

A visit to Crathes Castle in Scotland led to a change in direction in Helen's painting style and fostered a lifelong interest in gardens. She describes her style as impressionistic, favoring vibrant coloration and atmospheric effects. She also switched from oils to fast-drying acrylics. Yearly travels to Italy and the south of France have inspired both her painting and the design of her own garden. She cites the use of mirrors in her garden as something she gleaned from her travels.

Figure 9.21 Helen Lea's garden provided inspiration for many of her paintings. (Kilroy)

Helen has shared her talents as a member of WGC, designing the scarf for the 2005 GCA Annual Meeting and opening her garden for pop-up tours. She has received major awards at GCA flower shows in both photography and botanical jewelry. Members look forward to her annual painting shows in her Brookside home not far from where she got her start at her Border Star easel.

Laura Lee Carkener Grace

In many cases the exposure to gardens, especially those of family members, has an unforeseen influence on people later in life. Many garden club members cite the gardens of their mothers or grandmothers as having inspired their own creativity. Such was the case with Laura Lee Grace, whose grandmother had extensive "manicured gardens" in what is now Mission Hills, Kansas. On a scale rarely seen today, they included prize-winning peonies, gladiolas, and roses. No photographs exist of the gardens, and all the land has been subdivided.

Nevertheless, Laura Lee kept the vision of her grandmother's garden stored in her memory bank as she continued her education at Sunset Hill School and Wellesley College, where she majored in art history. As she put it, she "*studied*" art but never tried to "*do*" it until she met Kansas City area painter and teacher Donna Aldridge. Laura Lee attended Aldridge's classes for twenty years—two or three days a week for four hours at a time. For Laura Lee, the experiences she had mixing color under Donna's guidance were unforgettable, but she especially treasures her memories of Donna's teaching style.

Figure 9.22 The extensive gardens of Laura Lee Carkener Grace's grandmother inspired Laura Lee's exquisite flower portraits and the hat she wore (designed by Dody Gates Everist) for the 2005 Annual Meeting. (Dody Gates Everist)

Laura Lee mostly worked in oils, but sometimes she turned to the use of pastels. She considers herself as having been a realist in style. Although she often chose landscapes and animals as subjects, time and again she turned to the flowers she remembered from her grandmother's garden, newly seeing them through an artist's eye. Laura Lee's mentor, Donna, died recently, coinciding with Laura Lee's move to a smaller space, where she no longer paints. When asked, Laura modestly says, "All I ever was was a student." Many would disagree. ❧ 🌿

CHAPTER 10

"Everything IS Up-to-Date in Kansas City!"

THE WESTPORT GARDEN CLUB HOSTED its third zone meeting April 28–30, 1987. Members of twenty other garden clubs from seven states arrived in Kansas City for the meeting. WGC President Marie Bell O'Hara played hostess to Kay Donahue of Rye, New York, the president of GCA. Sallie Bet Watson and Sally Wood, co-chairs of the event, chose the zone meeting's theme: "Parks and Fountains." The delegates all stayed at the Alameda Plaza Hotel, where each guest received a Waterford vase (contributed by Adele Hall from Halls Kansas City department store) filled with lily of the valley and a gift-laden, bright yellow tote bag embossed with the club's dogwood emblem.

For the delegates, highlights of the 1987 meeting included touring and having dinner at the Nelson-Atkins Museum of Art; enjoying the lovely gardens of Margot Peet, Lillian Diveley, Betty Goodwin, and Marie Bell O'Hara as well as the arboretum at Linda Hall Library; and hearing great presentations by renowned conservationist Patrick F. Noonan, who gave the talk "Conservation Challenges: The Decade Ahead" and horticulturist Everitt L. Miller, director emeritus of Longwood Gardens near Kennett Square, Pennsylvania.

However, what many delegates thought was the most wonderful part of their stay was the opportunity to have dinner in the private homes of WGC members. Marilyn Pierson Patterson, Jean McDonald, Helen Sutherland, Barbara Seidlitz, and Norma Sutherland hosted cocktail-dinner parties, while Adele Hall and Dee Hughes were hostesses at a Kansas City Country Club luncheon. The zone meeting was so successful that the WGC vowed to follow the same format when GCA next came to Kansas City.

Figure 10.2 Delegates to the 1987 zone meeting toured four lovely gardens, including the garden of Margot Peet. (Gibson)

Figure 10.1 *Left:* The theme for the third zone meeting held in Kansas City was "Parks and Fountains." Delegates saw the fountain in the Laura Conyers Smith Municipal Rose Garden when they visited Loose Park. (Kilroy)

The club was scheduled to host its fourth zone meeting in 2003. Marion Enggas Kreamer, the very astute treasurer, determined that this time around, the club needed to start planning and raising money for the meeting early—even ten years before the event! In April 1992, the WGC Finance Committee recommended holding four in-house fundraisers beginning in 1993 and occurring every three years: 1996, 1999, and 2002. The first of these member events took the form of a "Run for the Roses" Kentucky Derby party held at Susie Vawter's home, which raised about $1,600. A house and garden tour hosted by five WGC members occurred in spring 1999, with proceeds totaling around $2,000.[1]

However, some seven years into it, preparation for the 2003 zone meeting took a left turn when at the club's annual meeting on June 1, 1999, held at Powell Gardens, members learned of a change in plans. GCA now wanted the WGC to host the 2005, much larger GCA Annual Meeting instead of the 2003 zone meeting. Margaret Hall, the incoming Zone XI chair, presented the proposal. She had already talked with WGC member Adele Hall, who as a civic leader, philanthropist, and the wife of Hallmark Cards President Don Hall, had considerable influence in the city. Adele supported the idea and stated her position at the meeting when one senior WGC member expressed her concern, saying, "We can do zone meetings, but we could never do an annual [GCA] meeting." All in attendance hoped that the worried member would graciously "eat her words" after the women of the WGC hosted a successful GCA Annual Meeting in 2005. However, in order for them to make that meeting a success, Margaret Hall advised members that they would need to block out everything else on their calendars for 2005, adding that "no excuses would be accepted except death."[2]

Years of Preparation

Naturally, hosting a GCA annual meeting would require much more time and effort than would holding a zone meeting. Although all eighteen clubs in Zone XI took charge of one or more aspects of the meeting, as the host club the WGC bore the lion's share of the responsibility. WGC members would have to secure the hotel (they ended up going with the Westin at Crown Center); plan all the side trips, tours, and dinners; arrange transportation; set up the GCA boutique; determine the schedules for the photography show, the flower show, and the plant exchange; and, in case anyone got into a tangle, get legal contracts drawn up. For this pro bono legal work, thanks went to the men's auxiliary, in this case John Readey and Jim Mathews. Margaret Hall and Betty Goodwin were named co-chairs of the event along with Diane McGauran of Milwaukee. Each of seven honorary chairs, all from Zone XI, including WGC's Norma Sutherland, had served at one time as a GCA vice president.

First, members of the steering committee, which included GCA members with previous annual meeting experience, determined the theme for the annual meeting. The idea of "Exploring" came from Mary Stanley, a member of the St. Paul Garden Club. While many GCA members would be exploring Kansas City for the first time, the theme also reflected the fact that many migrants, mountain men, explorers, and homesteaders had passed through the Zone XI states on their westward journeys. The trailheads of the Oregon and California trails were both in Independence, Missouri; the Santa Fe Trail, in the settlement of Westport.

Early in the planning stage for the annual meeting, the steering committee also decided on the sunflower (*Helianthus annuus*) as a logo for the meeting. Not only is the sunflower native to all six states in Zone XI, but also it is used for food products, thus representing an important agribusiness of the region. Karen Strohbeen of Des Moines and Betty Hubdyia of Wisconsin designed the sunflower logo that would be used on most everything associated with the meeting, including the tote bags that would go to all 650 delegates in attendance.

Each of the seventeen clubs from the six states in Zone XI provided gifts to put in the tote bags. When Susie Vawter volunteered to receive and store the gifts for all the delegates at her home, she assumed all the gifts would be small. However, one day an eighteen-wheeler pulled up in front of her house, and the driver wanted to know where to unload 650 boxes of cereal. Susie sent out a panicked call for help! With a little maneuvering, she and a rescue team of WGC members managed to fit the cumbersome and heavy boxes of cereal into Susie's carriage house.

WGC members also needed to determine their "uniforms" for the meeting. The cherry red smocks were long gone, but the needlepointed dogwood

Figure 10.3 *Right:* The annual meeting tour of Mission Hills, Kansas, included the Sunken Garden, which is in one of the city's fifty-seven parklets. (Kilroy)

Figure 10.4 "Exploring" was the theme for the 2005 GCA Annual Meeting, and the motif was the sunflower (*Helianthus annuus*).

pins remained a necessary accessory. At the May 2004 meeting, President Paget Higgins announced, "If you don't have a dogwood pin, go to The Studio *now*." The Studio was the needlepoint shop that offered dogwood-printed canvases for sale, and with less than a year before the annual meeting, members without dogwood pins needed to get busy stitching or hire someone to do it for them. Khaki pants, a white shirt, and the annual meeting scarf made up the rest of the uniform.[3]

The Scarf Is Its Own Work of Art

One obligation of the host club is to produce an annual meeting scarf. The first GCA scarf, sold at the 1993 Annual Meeting in Chicago, had not been billed as an "annual meeting scarf." The scarf design originated as a watercolor painted by botanical artist Cherie Sutton Pettit for the GCA Conservation Committee. Her painting, which showed the world encircled by images of endangered plants, hung in the GCA headquarters in New York. Inspired by the beautiful design, staffers came up with the idea of reproducing Cherie's image onto a scarf, which they could sell at the annual meeting to raise funds for the GCA Scholarship Fund.

The first scarf actually designated as an "annual meeting scarf" appeared for sale five years later, at the 1998 Annual Meeting in Williamsburg, Virginia, hosted by Zone VII. Designed by Frankie Welch, this scarf featured flowers representing the four states in Zone VII. Thus, a tradition was born. In 1999, GCA returned to Cherie Pettit for the design of the annual meeting scarf. Her rendering featured white lilies, stewartia, dogwood, and kalmia on a black background.

At the 2005 Annual Meeting in Kansas City, Cherie Sutton Pettit received the GCA Eloise Payne Luquer Medal "for special achievement in botany which may include medical research, the fine arts, or education." She was praised as "an artist with exceptional ability and generosity" and recognized for her many contributions to GCA besides her talent for creating beautiful scarves. When she passed away in 2008 at age eighty-eight, her obituary noted, "She was incredibly proud of her contributions to the GCA."[4]

For the annual meeting scarf in 2005, Co-Chairs Margaret Hall and Betty Goodwin turned to their fellow club member Helen Lea, who was not only an artist but had worked as a fabric designer. She did not take on this project lightly; she felt an obligation to design and produce the best scarf possible and began by researching how annual meeting scarves had been done in the past. She learned that GCA clubs had used the services of a young woman in the DC area who represented a French company, Cédric Brochier Soieries. Helen flew to DC and met with the representative, who suggested that Helen deal directly with the owner of the silk company in Lyon, France. Helen collaborated online with him, and together they came up with three color ways for Helen's design, "Sunflowers," in keeping with the meeting's motif. However, when the actual scarf samples arrived, Helen said "they were awful." As time was running out, she decided: "Nothing could be done except I go to France and oversee the screens and colors." Betty Goodwin offered to go with her. They bought the cheapest tickets they could find, which meant staying over a Sunday night (the club did not pay for their trip), and set out on what Helen later called "an adventure of a lifetime." The owner of Cédric Brochier Soieries met them at their hotel in the historic silk district and escorted them to the factory, where they were able to mix the colors themselves and after three screen passes, achieve the perfect combination. After a wonderful week in France, Helen and Betty returned to Kansas City with a beautiful silk-screened scarf done the old way on good silk with hand-stitched hemming. They proudly showed off the scarf at the October 2004 WGC meeting. Their only fear concerned the pricing of the scarf at one hundred dollars, but the scarves sold out and made a modest profit. Past and current WGC members consider the *Helianthus* scarf as one of the most beautiful of the many annual meeting scarves they have seen. Still today, a new scarf is produced for each annual meeting and sold in the Market Place on the GCA website.[5]

Garden Tours Show Off the Beauty of the Area

Of course, WGC members wanted to make sure the many out-of-town delegates to the 2005 GCA Annual Meeting visited beautiful gardens in the Kansas City area. In fall 2004, Marilyn Hebenstreit organized WGC volunteers to plant a thousand daffodil bulbs at the Linda Hall Library gates and six hundred tulips at the library's entrance. The Transportation Committee planned a scenic bus route for delegates to take on their way to view private gardens in Kansas City, Missouri; Prairie Village, Kansas; and Mission Hills, Kansas. When Lorelei Gibson explained to the Mission Hills Park Board that hundreds of out-of-town gardeners would be touring Mission Hills, the city agreed to plant tulips and pansies to make the tour of Mission Hills especially colorful.

The 2.1-square-mile city of Mission Hills is unique in that it includes no businesses or industries—only single-family homes and three golf courses: Mission Hills Country Club, built around 1915; The Kansas City Country Club, relocated to Mission Hills in 1926; and Indian Hills Country Club, on the south edge of the residential area, opened in 1927. Part of J. C. Nichols's original plan was to buffer the homes with golf courses to avoid encroachment of less desirable development. Annual meeting delegates

Figure 10.5 Helen Lea designed the beautiful "Sunflowers" annual meeting scarf; the project even involved a trip to France.

were duly impressed by Mission Hills, as were the members of the GCA Awards Committee. After the annual meeting, Lorelei Gibson wrote the application that the WGC successfully submitted for a GCA Club Civic Improvement Commendation for the City of Mission Hills. In June 2007, Lorelei Gibson herself received the GCA Club Civic Improvement Award "for many years of dedicated leadership for the Mission Hills Park Board and your deep knowledge of plant material to beautify and preserve the city."[6]

Lorelei Manning Gibson: Horticulture and Historic Preservation

Lorelei Gibson, a Westport Garden Club member for more than thirty years, enthusiastically embraces all facets of the club's interests—including garden and floral design, photography, horticulture, and historic preservation.

Figure 10.6 One of Lorelei Gibson's talents lies in the area of floral design as she enhances her own home with beautiful arrangements. (Photo courtesy of Lorelei Gibson)

Although both of Lorelei's grandmothers were gardeners, her own garden awakening did not begin until she and her husband, David, and their blended family moved into a large house in Mission Hills, Kansas, one with historic associations. Formerly the home of the Samuel Sosland family, the house incorporated large stones from the William Rockhill Nelson home in Kansas City, Missouri, as well as a basement full of roof tiles from that same residence. Finding herself surrounded by rambunctious teenagers, Lorelei sought respite outdoors in the old house's neglected garden. She discovered a series of deep ponds. As she worked to restore them, much to her delight

Figure 10.7 *Left:* Lorelei Gibson has worked to maintain the beauty of Mission Hills, Kansas, as chair of its Park Board. (Kilroy)

dormant lotus seeds came to life, and the ponds became a defining feature of the garden.

Lorelei has since changed residences twice, restoring and creating gardens at both new locations. These two gardens are documented in the Smithsonian Archives of American Gardens. At the 2005 GCA Annual Meeting, delegates toured one of these gardens, which had previously been owned by another Westport Garden Club member, Margot Peet. Lorelei also has a garden in Santa Fe, New Mexico, which has been on the Santa Fe Garden Club's "Behind Adobe Walls" tour.

Lorelei is a member of the Mission Hills Park Board and has served as its president for more than twenty-five years. The bylaws of this board still include the provision that one member must be from The Westport Garden Club. The board has the final say on aesthetic decisions regarding plantings and sculpture in the City of Mission Hills, Kansas.

Lorelei's advice to new gardeners is to keep a garden diary with photos. Her own photo images raise the bar considerably, since she has been the recipient of a GCA Certificate of Excellence in Photography.

Lorelei has also been recognized with the GCA Club Horticulture Award and the GCA Club Civic Improvement Award. ∽

Plant Exchanges Broaden Everyone's Knowledge

Annual GCA meetings and some zone meetings used to feature a judged plant exchange (PX). The purpose of the PX was to provide hands-on horticultural education for all members by fostering experimentation with different methods of propagation, discovery of new plant materials, learning about landscape uses for many kinds of plants, identification of plant species, and sharing plants with other GCA members across the country. Traditionally, each club brought six plants to the PX and took six home. All clubs used the same style of container—a twelve-ounce cup—and a "soil-less" mixture to avoid exchanging plant diseases. Judges took into account the quality of the specimen, the appropriateness of the selection of that plant type, the difficulty of propagation, and the accuracy of the entry card.

Norma Sutherland wrote a column for the October 1981 issue of the GCA *Bulletin* titled "Those Plant Exchange Babies." In it, she asked, "Do you ever wonder about the whereabouts and the welfare of your babies that you have sent forth for adoption at the GCA Plant Exchange?" Norma had heard from a woman in High Falls, New York, who had adopted her little *Iris cristata* at the exchange three years before. The plant had prospered and multiplied, something Norma found highly satisfying. Now she urged all adopters of exchanged plants to report on their adoptees.[7]

The WGC minutes do not mention the GCA national plant exchange until 1975; they mention it again in 1979, and then in 1981, when the WGC received a GCA Plant Exchange Merit Award at the annual meeting in Cincinnati. In 1984, four members—Norma Sutherland, Susie Vawter, Betty Goodwin, and Marie Bell O'Hara—brought home plant exchange awards from the annual meeting. When Susie Vawter was WGC Horticulture chair in 1987, she reported that WGC was awarded a GCA Horticulture Committee Commendation for six specimens propagated by six different members—wild ginger, grown by Jean McDonald; big blue lobelia, by Norma Sutherland; variegated Solomon's seal, by Diana James; lady's mantle, by Betty Goodwin; sedum, by Marie Bell O'Hara; and foamflower, by Susie Vawter.

Perhaps the best story about the plant exchanges dates from 1988, when Sallie Bet Watson served as WGC president. At the time, she and Betty Robinson asked DeeDee Adams to accompany them to the GCA Seventy-Fifth Anniversary Annual Meeting in Detroit. DeeDee was a new member, having been initiated in 1985, and so was a bit intimidated about going to the annual meeting in the company of these very experienced older women. She recalls sitting on the airplane in between Sallie Bet and Betty, holding in her lap the six plants destined for the plant exchange. It was 7:00 a.m., the women having decided to take an early flight. Looking down at the plants in her lap, DeeDee suddenly realized they had tiny white dots on them—and the dots were moving! In distress, she informed her seatmates. "No problem," said Sallie Bet as she called over the stewardess and ordered three Bloody Marys. Now DeeDee went from distressed to aghast. She had left two small children at home, and she was on a plane to Detroit drinking Bloody Marys at 7:00 in the morning. When the drinks arrived, Sallie Bet and Betty advised DeeDee to use her vodka on the plants. "Just rub it on those little white things, and they will be gone," they told her. Obviously, the older women had learned the

ropes, the WGC winning six Awards of Merit for the (pest-free) plants they exchanged at the meeting.[8]

Preparing for a plant exchange struck as much fear in the hearts of many a club member as did the idea of entering an arrangement in a flower show. In 1993, Horticulture Committee Co-Chairs Kathy Gates and Kelly Lambert issued a word of warning to incoming committee members: "When you hear the words: zone meeting, regional meeting or conference, plant exchange or any gathering of garden clubs taking place outside the city, start asking questions because this probably means you." The role of the Horticulture Committee is to plan one or two horticultural programs each year, to give horticultural tips at each meeting, and "to gather/grow, prepare and deliver plants for the plant exchange."[9]

In June 1993, the same month and year that Kathy Gates and Kelly Lambert issued their warning about plant exchanges, Marie Bell Watson O'Hara received the GCA Club Horticulture Award, based on her extensive garden and her many contributions to the WGC, including propagating plants for the PX. In 2004, WGC members won six blue ribbons in the plant exchange. Joy Jones and Gina Miller took home first place ribbons, while Norma Sutherland and Margaret Hall each received TWO first place ribbons.[10]

The theme for the 2005 Annual Meeting Plant Exchange was "Right Plant, Right Place." Each club selected one environmental condition—dry, moist, wind, salt, or high altitude—to emphasize what could be grown in a specific location. WGC members had propagated hardwood cuttings for the exchange, with the rootings being nursed in Margaret Hall's garage until time for the show.

As it turned out, the 2005 PX would be the second to last such event. At the annual meeting in Kansas City, Barbara B. Shea, the head of the GCA Horticulture Committee, announced the cancellation of plant exchanges after the 2006 Annual Meeting "due to growing concern about the risk of spreading disease." Obviously little white moving things on plants were not unique to our club, despite care, attention, and the application of vodka. Over the years, approximately 38,500 plants were exchanged.[11]

Figure 10.8 *Left:* Although there are more than 300 species of irises, including this popular bearded variety, Norma Sutherland contributed her woodland *Iris cristata* to the plant exchange and heard from the woman who had adopted it.

It's Showtime: April 22–27, 2005

The annual meeting actually began on Wednesday, April 20, 2005, with the early arrival of the members of the Horticulture Committee only. That evening a cocktail party was planned for them at the apartment of Joan "Jody" Jenkins and Paul Dana Bartlett, but the Hort Committee delegates arrived late. As they were preparing to leave the hotel to go to the party, they were surprised to look out the window to see pool chairs being removed. Hotel staff informed them that there was a tornado alert; after all, they were in *Kansas* City, part of *Wizard of Oz* territory, even though the hotel was actually in the state of Missouri. For their part, WGC members did not think the weather looked very threatening by KC standards, and they encouraged the Hort delegates to attend the party, which they eventually did.

On Thursday, April 21, the Hort Committee members enjoyed a tour organized by seedling WGC member Ginny McCanse, which included a trip to Linda Hall Library. It is a tradition at GCA annual meetings that the Horticulture Committee commemorates the occasion with the gift of a tree to the host city. The gift tree, planted in the arboretum on the library grounds, was the *Acer saccharum* 'Green Mountain,' commonly called a sugar maple.

On Friday, early arriving delegates enjoyed the option of a series of pre-tours, and that evening, April 22, Kansas City Mayor Kay Barnes officially kicked off "Exploring." The meeting featured outstanding speakers. Dr. Peter Raven, botanist, environmentalist, and longtime director of the Missouri Botanical Garden, gave the keynote address. Respected northwest gardener and author Ann Lovejoy, often identified as "the Mother Theresa of organic gardening," spoke at the horticulture meeting.

On Saturday, April 23, delegates registered for the PX. Meanwhile, the *Helianthus* Boutique, special exhibits, and photography show opened in the hotel. Helen Lea, in charge of the photography show, had been overwhelmed by the deluge of photographs that arrived at her home. Thanks to help from Adele Hall, all the photographs were mounted, properly lit, and hung at the hotel for the opening of the show on Saturday. Delegates served as judges of the photography show, as did passersby; all could vote for their favorite photograph in the exhibition called *Exploring Gardens, Plants, and Water*.

The first official evening of the meeting, all delegates attended zone dinners, held at eleven different restaurants. On day 2, Sunday, April 24, delegates could opt to participate in various workshops or programs or take tours of Hallmark's Fine Art and Photograph Collection, Linda Hall Library, the Toy and Miniature Museum, the Nelson-Atkins Museum of Art, the Steamship Arabia, the Truman Library, or Trapp and Company. The Kemper Museum was the venue for Sunday night's dinner and was also the location of the floral exhibition.

On Monday and Tuesday afternoons, April 25 and 26, half of the delegates (Group A) had lunch at Union Station and then toured the many fountains in the City of Fountains, Kauffman Garden, and four private gardens—the gardens of Dody and Lathrop Gates, Jan and Ned Riss, Lorelei and David Gibson, and Margaret and Tom Hall. The other half (Group B) visited Powell Gardens, with the reverse happening the following day. While at Powell Gardens, delegates were offered a selection of take-home gift plants: bottlebrush buckeye (*Aesculus parviflora*), sweet spire (*Itea virginica* 'Henry's Garnet'), dwarf Korean lilac (*Syringa meyerii* 'Palibin'), or doublefile virburnum (*Viburnum plicatum var. tomentosum* 'Summer Snowflake').

On Monday evening, April 25, the members of The Westport Garden Club actually fulfilled their self-made vow to entertain GCA members in their own homes. In a remarkable display of generosity and organization, thirty-five WGC members—Jody Bartlett, Jeanne Bleakley, Wendy Byers Brinton, Lyndon Chamberlain, Dody Gates, Lorelei Gibson, Betty Goodwin, Adele Hall, Margaret Hall, Marilyn Hebenstreit, Paget Higgins, Ellen Hockaday, Blair Hyde, Alison Jager, Joy Jones, Nancy Lee Kemper, Betty Kessinger, Flip Kline, Carolyn Kroh, Kelly Lambert, Ginny McCanse, Martha "Mimi" Cate Newell, Marie Bell O'Hara, Sharon Orr, Ginger Owen, Marilyn Patterson, Susan Pierson, Wendy Powell, Ann Readey, Linda Spencer, Catherine "Cacki" Smith, Alison Ward, Sally West, Sally Wood, and Eulalie Zimmer—held dinner parties in their homes with co-hostesses from other clubs in Zone XI. Everyone had the same meal: Chicken Maciel, rice, roasted vegetables, and Prospect salad. Ridge Watson, vintner and son of WGC member Sallie Bet Watson, provided wine at cost.

For both the private garden tours and the dinner parties, the WGC had made a monumental effort to get permission to drive buses into Mission Hills. That permission granted, on the night of the dinner, the bus driver dropping off delegates at Eulalie Zimmer's home did not want to take the bus up the steep driveway. He decided to drop his passengers off as close

Figure 10.9 Lyndon Chamberlain agreed to be one of the hostesses for the annual meeting, but her mother, WGC member Peggy Gustin, told her she needed a "better house" to host a GCA dinner. Her grandmother, founding member Hester Milligan Gustin, would never approve of her current home. Therefore, Lyndon and Rick found a "better house," and Mrs. Gustin bought them new placemats and napkins to use for the GCA dinner. (Chamberlain)

to the curb as he could get. In so doing, he managed to get the bus stuck, damaging the bus steps! While the passengers managed to disembark, disgruntled neighbors called the police to complain about the bus. When the police arrived, they insisted on interviewing the dinner guests as potential witnesses, asking them not only their names but their AGES! All agreed that Eulalie's dinner party was one not to forget! The delegates, unabashed, worried about the fate of the bus driver; all of them wrote letters to the bus company saying that he had not been at fault but was only trying to be helpful. Such is the courtesy of GCA members!

Figure 10.10 *Left:* Blair Hyde was one of the thirty-five WGC members who volunteered to host a dinner and show off her garden for the annual meeting. (Hyde)

The City of Fountains awards dinner, held on Tuesday night, April 26, 2005, honored ten individuals, including Claudia Alta "Lady Bird" Johnson, who was unable to attend. At the dinner, the GCA gave sunflower arrangements in sparkly red Dorothy shoes to VIP guests. One of those guests was Ford Motor Company President and CEO William Clay Ford Jr., who was there to receive the Frances K. Hutchinson Medal, awarded to the Ford Motor Company "for distinguished service in conservation." With the mission "Reduce—Reuse—Recycle," the company had reduced the amount of water used in the manufacture of their automobiles by over 5 billion gallons, reused 225 million pounds of paint solvent, recycled 50 million plastic soda bottles to make vehicle parts and 1 billion bottle caps to make heating and air conditioning components, and ensured that each vehicle that Ford built was 85 percent recyclable. Being awarded a GCA medal is the highest of honors, and Bill Ford was happy to receive the recognition for the Ford Motor Company. In his acceptance speech, he said that his mother had told him he must be on his best behavior because it was a garden club event.[12]

When comments came in about the annual meeting, the delegates raved about the "fabulous" City of Fountains banquet, the tours of private gardens, and the "outstanding" dinners in WGC members' homes. Delegates also offered many words of praise about the optional tours. However, the post-meeting trip on Wednesday, April 27, to the Flint Hills of Kansas created the biggest "Wow!" The more adventurous in the group walked among wildflowers and other native flora and, as a special treat, watched a planned prairie burn.[13]

In October 2004, the GCA *Bulletin* chair had asked for each club to vote on whether to continue to publish an extra issue to cover the 2005 Annual Meeting. Since the WGC was hosting the event, WGC members voted a definitive "yes," as did the members of most other GCA clubs. The resulting twenty-four-page *Bulletin* was subtitled "Everything IS up-to-date in Kansas City! From the incredible skyline . . . to well-labeled buses . . . and fountains too beautiful and too numerous to count."[14]

While, as mentioned, WGC members had raised funds over the course of several years in anticipation of hosting the next GCA zone meeting, when the national organization had them hold the annual meeting instead, they found that the national budget covered many of the expenses. Everyone was

Figure 10.11 GCA published a special twenty-four page *Bulletin* about the 2005 Annual Meeting with the headline "Everything IS Up-to-Date in Kansas City!"

After hosting three zone meetings and an annual meeting in Kansas City, it is fitting indeed that in 2025, in the same year The Westport Garden Club will celebrate its seventy-fifth anniversary, the club will again host a Zone XI meeting, bringing the club full circle back to its beginnings and rehearsing its history. 🌿

thrilled to learn that the 2005 GCA Annual Meeting netted $150,219.64. After the WGC returned all the seed money (with interest) to all the Zone XI clubs and repaid contributions made by the GCA Board of Associates, $46,428.52 remained, half of which went back into the GCA endowment. For its part, the WGC ended up with $23,214.26 to donate as it wished. The WGC gave $10,000 to Powell Gardens and $2,000 to the Missouri Botanical Garden by way of thanks to Peter Raven for his presentation. The club also used the proceeds from the annual meeting to make other gifts to community organizations that had gone above and beyond to help them with the event, including Kansas City Community Gardens, the Toy and Miniature Museum, the Noteables, and Pembroke Hill School.[15]

Figure 10.12 *Left:* The annual meeting post-trip to the Flint Hills of Kansas created the biggest "Wow!" (Sink)

CHAPTER 11

Awards: Treasured Recognition for Outstanding Work

THE GARDEN CLUB OF AMERICA has a commitment to excellence; therefore, "a key priority of the GCA is to recognize clubs, club members, and non-members whose accomplishments and efforts further the GCA purpose." Special award committees at the club, zone, and national level oversee the selection of worthy individuals and organizations. In 1917, GCA created its first national award, and although it has established many other awards since then, becoming a GCA national medalist has always been deemed the greatest of achievements.[1]

National Medalists and the WGC

Notably, the WGC has received a national medal and has proposed three individuals who went on to become national medalists. In 1964, the GCA awarded The Westport Garden Club itself the Amy Angell Collier Montague Medal for the club's "vision, its energetic and continuing advisory, manual, and financial help in making the Junior Gallery and Creative Arts Center of the Nelson Gallery and Mary Atkins Museum of Art in Kansas City, Missouri, an important part of the cultural growth of the community."[2]

The first of the three national medal winners proposed by the WGC was Anita B. Gorman. In 2010, Gorman received the Margaret Douglas Medal from the GCA "for notable service to the cause of conservation education." In the early 1960s Anita Gorman led the fight to save a Native American archaeological site, and she protected open land in North Kansas

Figure 11.1 *Left:* Azaleas bloom in Kristie Wolferman's garden. (Wolferman)

City from development, the site later becoming the Anita B. Gorman Park. A founding member of the City of Fountains Foundation, she became the first woman appointed to the board of the Kansas City, Missouri, Parks and Recreation Department, serving as its president from 1986 to 1991. In 2005, the Missouri Department of Conservation named its conservation center the Anita B. Gorman Discovery Center in recognition of her support of many conservation programs. Before Anita Gorman received the GCA medal, the WGC had successfully nominated her in 2008 for the GCA Zone XI Conservation Commendation. She dedicated her life (1930–2023) to conservation and she won many awards, but becoming a 2010 GCA National Medalist is among her highest tributes.

In 2012, the GCA awarded Dr. Wes Jackson, co-founder and executive director of The Land Institute in Salina, Kansas, the Elizabeth Craig Weaver Proctor Medal "for exemplary service and creative vision in any field related to The Garden Club of America's special interests." Dr. Frederick Kirschenmann, who wrote on behalf of Dr. Jackson, noted Jackson's continuing contribution to "the restoration of the biological health of our soil and biodiversity of our landscape—an achievable goal with well managed perennial crops." Wes Jackson's quest to develop food-production methods that sustain the land and soil cannot be achieved easily, but as he put it, "If your life's work can be accomplished in your lifetime, you are not thinking big enough."[3]

Then, in 2023, GCA honored Robert J. "Bob" Berkebile with the Cynthia Pratt Laughlin Medal "for outstanding achievement in environmental

Figure 11.2 A key priority of GCA is to recognize club members' accomplishments. Laura Sutherland, Cindy Cowherd, Nancy Lee Kemper, and Margaret Hall attended a Zone XI meeting where (*in front*) Lyndon Chamberlain and Kathy Gates won awards. (WGCA)

Figure 11.3 GCA honored WGC honorary member Bob Berkebile with the Cynthia Pratt Laughlin Medal at the annual meeting in 2023. GCA President Debbie Oliver and Awards Chair Jana Dowds presented him with a framed certificate. (Mary Ann Powell)

protection and the maintenance of the quality of life." This medal followed the GCA Zone XI Conservation Commendation he had received in 2021, also proposed by the WGC. An internationally known "green" architect, Berkebile advocates for sustainable designs that respect the environment and enhance the community. He helped launch the US Green Building Council and has helped restore and revitalize communities devastated by disaster or blight. In 2009 Bob Berkebile became an honorary member of the WGC and, with Kathy Gates, led the charge to form the organization now called Deep Roots KC. Berkebile's current work with rehabilitation of the Blue River watershed is supported by WGC members. As a civic leader and environmentalist, Bob Berkebile reminds us daily that "the future of the Earth is in our hands."[4]

Another national award that has special meaning to WGC members is the Elizabeth Abernathy Hull Scholarship Award, for not only did Hull have Kansas City roots, but it was she who inspired the club's formation.

The award she established recognizes "individuals across the country, who through working with children under sixteen years of age in horticulture and the environment, have inspired their appreciation of the beauty and fragility of our planet." The WGC has successfully sponsored four individuals for the Elizabeth Abernathy Hull Award: William Sappof in 1994; Barbara Plapp in 2003; Barbara Montague in 2007; and Andrea Salisbury in 2009.[5]

GCA Zone Awards and Commendations

Apart from national medals, which represent the highest of achievements, awards are given on the zone and club levels as well. While only GCA members are eligible to receive "awards," nonmembers can be given "medals" and "commendations." Nine WGC members have received a total of eleven GCA Zone XI awards; both Norma Sutherland and Margaret Hall have received two.

The most recent WGC zone award recipient is Laura Babcock Sutherland, who served as club president from 2015 to 2017 and who is the co-chair of the 2025 zone meeting. Laura received the GCA Zone XI Communications Award "for utilizing effective communications to promote the purpose of The Garden Club of America." This award was announced at the Zone XI meeting on September 20, 2023, in Chicago and was celebrated again at the October 2 WGC meeting. The citation on Laura's award reads: "In recognition of her journalism talents and gracious enthusiasm given generously to produce countless GCA communications and resources, connecting members across zones." Laura had previously received the GCA Club Communications Award, in 2018.[6]

The GCA's highest zone award is the Creative Leadership Award (CLA). The CLA goes to a GCA club member "for outstanding and sustained leadership skills supporting the purpose of The Garden Club of America." The GCA president always presents the CLA at the respective zone meeting. Often years pass without this award being proffered in any given zone, for the recipient needs to be approved at the zone level, have six letters of recommendation (with only the proposing letter from the candidate's own club), and be known to and approved by every member of the GCA Executive Committee. Many clubs have never had a CLA recipient, as such a recipient not only is recognized for her leadership skills but also must have extensive GCA experience at the club, zone, and national levels. Thus, the WGC is deeply honored to have had three members receive the GCA Creative Leadership Award: Julia Tinsman was so honored in 1982 at the zone meeting in St. Louis, Norma Sutherland in 1988, and Margaret Hall in 2007. Sally West, who was serving as the WGC president when Margaret Hall received the CLA, flew up to Wisconsin to accompany Margaret for the presentation—this even though Sally's daughter was getting married in two days! Such is the importance of this award.[7]

Margaret Weatherly Hall: Contagious Enthusiasm

For many members of The Westport Garden Club, the name Margaret Hall is all but synonymous with the club itself. While other leaders have helped guide it, no one has done as much as Margaret to preserve the club's history and help new members gain a better understanding of GCA. "Ask Margaret!" is a common answer to many a question!

When Margaret joined The Westport Garden Club in 1988, she was a traditional flower arranger. However, a class in ikebana opened her eyes to a floral design of simplicity and creativity and started her on a journey of discovery. She delved into contemporary techniques and entered flower shows at all levels, never afraid to take chances.

Within about ten years she had received the GCA Club Floral Design Achievement Award "for her contagious pleasure in matters to do with flowers."

Margaret's enthusiasm also extends to horticulture, through which she discovered the miracle of propagation, and to photography, which provided her with a new way to see nature's quirky qualities. She became a judge in both photography and floral design, becoming so proficient that in 2011 she was chosen as a judge for the World Association of Floral Artists the only time the association has met

Figure 11.4 Walking stick in hand, Margaret Hall found rest, but only momentarily, on a butterfly chair. (Photo courtesy of Margaret Hall)

in the United States. Margaret is also a member of the Rare Plant Group, hosting its meeting in 2016.

Challenged by GCA, Margaret has taken on both zone and national offices, serving as Zone XI chair, representing Zone XI as a nominating representative, chairing the GCA Admissions Committee, and serving as a member of the GCA Executive and Stewardship Committees, among other positions. Overcoming her fear of speaking to large crowds, Margaret co-chaired the 2005 GCA Annual Meeting in Kansas City. Her list of accomplishments and recognition sound like a who's who of the GCA.

A lady with many GCA annual meeting scarves—she has twenty-five and counting—Margaret's contagious enthusiasm has transformed The Westport Garden Club. A former teacher, she has used her skills to mentor new members, and her garage is always open for horticulture experiments; her garden, for tours. There is literally no part of the WGC of which she has not been a part—from serving as president, to recording bylaws and membership, to chronicling history, as well as doing all things floral.

Figure 11.5 Club committee chairs gathered before an early fall meeting to show off Margaret Hall's vast collection of annual meeting scarves. (Roy Inman)

However, it is the experiences and friendships she values the most. As she put it, "It was an exciting day in 1988 when I received my invitation to join The Westport Garden Club. I will always treasure the friendships I have made and the many new skills I have learned." ✑

GCA Zone XI commendations have gone to four nonmembers. In addition to Anita Gorman in 2008, Doug Ladd, director of conservation for The Nature Conservancy of Missouri, received the Zone XI Conservation Commendation in 2019. As a conservation biologist, Ladd has made a great impact on Missouri's conservation efforts, and his dynamic presentations have had considerable influence not only on the members of the WGC but on all residents of the area, encouraging everyone to take care of our remnant prairies and our native plants.

For "extraordinary expertise in landscape architecture, garden design, plant accessions, horticulture education and authorship, and passion for the natural environment," Alan Branhagen received the GCA Zone XI Horticulture Commendation at the Omaha, Nebraska, meeting in May 2013. He was celebrated again in Kansas City at the WGC Annual Meeting on June 5, 2013, at Lyndon Chamberlain's home. As director of horticulture for Powell Gardens, Branhagen contributed significantly to both the Gardens and the WGC. When he left Powell Gardens, he assumed a new position in Minnesota and then, in July 2023, became the executive director of the Natural Land Institute in Rockford, Illinois. However, Branhagen continues to help the WGC in its plans to renovate an azalea garden in Loose Park. His influence is also felt through his publications, which are in many WGC members' libraries: *Native Plants of the Midwest: A Comprehensive Guide to the Best 500 Species for the Garden*, published in 2016, and his 2020 *The Midwest Native Plant Primer: 250 Plants for an Earth-Friendly Garden*.[8]

Recognized in an entirely different field, but one of great importance to GCA, was James B. Nutter, who in 2008 received a GCA Zone XI Historic Preservation Commendation for his part in restoring an historic neighborhood, subsequently called "Nutterville." His citation read: "For his love of the historic Westport neighborhood and his vision to preserve its very special homes and buildings for future generations to enjoy." Jim Nutter was a mortgage-lending pioneer whose company helped tens of thousands of families—including veterans, ethnic minorities, women, and historically underserved borrowers—to realize the American dream of owning their own homes. Located in the Westport neighborhood, where early pioneers once outfitted their wagons for treacherous journeys west (and the source of the name for The Westport Garden Club), James B. Nutter and Company opened in 1951 and stayed in business at the same location until 2022. Around 1975, Nutter came to the realization that many of the late nineteenth-century houses in the neighborhood surrounding his business lay in disrepair. Having a deep respect for the neighborhood's history, he began purchasing and restoring some of these old Victorian-era homes, giving them "painted lady" colors typical of the nineteenth century. The Nathan Scarritt house, one such home he restored, opened to the public in 2004. Built in 1847, it is the oldest surviving residence in either

Westport or Kansas City and is on the National Register of Historic Places. Coincidentally, Nathan Scarritt was the great-great-grandfather of WGC member Ann Readey, who was serving as club president when Jim Nutter received the GCA commendation.[9]

Jim Nutter, who believed in "the integral relationship between flowers, good homes, and good neighbors," was recognized not only for restoring more than thirty homes in Nutterville but also for landscaping them. At James B. Nutter and Company, Nutter said, "flowers are a passion. To us, planting flowers is one of the important ways to turn a house into a home." In 2006, the landscape company that Nutter hired, Signature Landscape, Inc., won the prestigious Grand Award for Commercial Landscape Maintenance given by the Mid-America Green Industry Council. When James B. Nutter died in 2017, the end of his obituary read: "In keeping with Jim's colorful character, civic spirit, and love of that place called home, thousands of flowers will continue to bloom each year in his memory in an area of old Westport that will forever be known as 'Nutterville.'"[10]

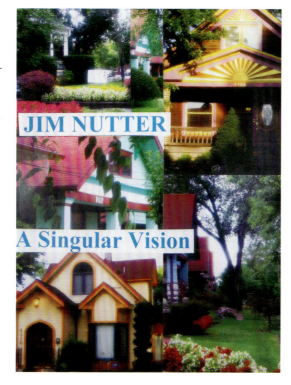

Figure 11.6 Jim Nutter had a singular vision to restore an historic Westport neighborhood, which came to be known as Nutterville. (WGCA)

Photography Adds to the Roster of Awards

The GCA Photography Committee, established in 2006 as a subcommittee of the Flower Show Committee, "furthers the knowledge and love of gardening and the natural landscape through photography." Although a new GCA field, photography has sparked the interest and enthusiasm of several WGC members, who enjoy organizing exhibitions, hosting seminars, and even publishing an informative how-to photography guide, *Think Before You Click*.

The first and, to date, the only GCA medal in photography (the GCA J. Sherwood Chalmers Medal) went to Michael Forsberg of Lincoln, Nebraska, in 2020. (WGC members had visited Forsberg in 2017.) WGC members have won club and zone photography awards, which "may be presented annually to one or more GCA club members for outstanding achievement in photography within and beyond the activities of the recipient's own club." Two WGC members, Margaret Hall and Lyndon Chamberlain, have received the GCA Zone XI Marilyn Sward Award for Photography. From the Garden Club of Evanston, Illinois, Marilyn Sward served on the first GCA Photography Committee and contributed in the winter of 2008 to the first edition of *Focus*, the GCA's online photography magazine. GCA club photography awards, which also may be presented annually to one or more GCA members, have been given to three WGC members: Betty Goodwin, in 2011, the author of *Think Before You Click*; Helen Lea, in 2017; and in 2022, Lyndon Chamberlain, "for her creative eye and passion for photography, generous mentorship of club members, her resourcefulness using GCA opportunities, and her award-winning recognition as a photographer." In addition, several WGC members have received photography awards at flower shows.[11]

Lyndon Gustin Chamberlain: Picture Perfect

Bubbly and enthusiastic, Lyndon Chamberlain has been a visual person since childhood. She remembers in great detail the "gorgeous" garden of her grandmother, Hester "Hettie" Milligan Gustin, a founding member of WGC. Located in Cotuit, Massachusetts, on Cape Cod, the garden included roses, sweeping borders, and a garden room dedicated to all white flowers. Lyndon's mother, Margaret "Peggy" Garner Gustin, who was also a WGC member, had a rose garden styled in a parterre fashion. However, it was not the family's gardening skills that influenced Lyndon to join The Westport Garden Club in 2004. Instead, she was drawn to the camaraderie of the WGC members and the GCA photography program, then in its infancy.

Figure 11.7 Many of Lyndon Chamberlain's winning photographs result from her wide travels with other GCA members. (Rick Chamberlain)

When Lyndon was appointed as co-chair of the 2011 GCA Flower Show, themed "Joie de Vivre," she had never been to a flower show, although she had successfully entered photographs in them. She quickly remedied the situation and never looked back. She is now an approved GCA photography judge and has served on the GCA Photography Committee and as coordinator of the Photography Study Group. The travel, the new friends, the workshops, and the shows have, as she puts it, taken her "out of the bubble" of her own club and into the larger world of GCA. She has learned new skills and continues to be impressed by the talented members of GCA. She encourages members to say yes if asked to be a delegate to annual meetings so that they,

too, may see the scope of GCA and perhaps get involved at the zone or national level.

Lyndon has been recognized with multiple ribbons at flower shows, with the WGC "Above and Beyond Award," the GCA Club Photography Award, and the GCA Zone XI Marilyn Sward Award for Photography. ❧

Special Awards Just for Club Members

Many clubs give their own special awards in addition to those bestowed by the GCA. The Westport Garden Club gives out four awards, not necessarily on an annual basis but when the club deems an individual worthy of receiving one. The WGC "Above and Beyond Award," first presented in 2003, recognizes extraordinary service to The Westport Garden Club. "The Deep-Rooted Perennial Award" can be given to a member who has demonstrated a deep devotion to the club and continues to give of her time and talent year after year. The Norma Henry Sutherland Award, founded in 2008 to honor Norma Sutherland, goes to a member of no more than five years' standing for "outstanding participation including attendance, volunteerism, and leadership."

To the casual observer, the plethora of awards may seem to undermine their importance, but those who are members of GCA clubs know that what they are doing to enhance the beauty of our country, conserve our resources, educate the public about horticulture and conservation, and protect our environment are worthy goals. Hard work is expected of all GCA and WGC club members, and those who make extraordinary contributions deserve recognition. 🌺

11.8 *Autumn Tulip.* Laura Lee Grace's flower portraits, here and throughout the following pages. (Laura Lee Grace)

Garden Club of America Awards
GCA National Awards

1964 GCA National Medalist

Amy Angell Collier Montague Medal
Recipient: The Westport Garden Club

2010 GCA National Medalist

Margaret Douglas Medal
Recipient: Anita B. Gorman (Mrs. Gerald W.)
Proposed by The Westport Garden Club

2012 GCA National Medalist

Elizabeth Craig Weaver Proctor Medal
Recipient: Dr. Wes Jackson, The Land Institute, Salina, Kansas
Proposed by The Westport Garden Club

2023 GCA National Medalist

Cynthia Pratt Laughlin Medal
Recipient: Robert J. "Bob" Berkebile
Proposed by The Westport Garden Club

GCA Zone Awards

GCA Zone XI Creative Leadership Award

1982 Julia Chandler Tinsman (Mrs. Charles Humbert)
1988 Norma Henry Sutherland (Mrs. Dwight Dierks)
2007 Margaret Weatherly Hall (Mrs. Thomas Bryan III)

GCA Zone XI Communications Award

2023 Laura Babcock Sutherland (Mrs. Todd Latta)

GCA Zone XI Conservation Award

1974 Betty Clapp Robinson (Mrs. Kip)
2002 Virginia "Gina" McDonald Miller
2018 Kathy Garrett Gates (Mrs. Kirkland Hayes)

GCA Zone XI Conservation Commendation

2008 Anita B. Gorman (Mrs. Gerald W.)
2019 Doug Ladd
2021 Robert J. "Bob" Berkebile, FAIA

GCA Zone XI Floral Design Achievement Award

2004 Margaret Weatherly Hall (Mrs. Thomas Bryan III)

GCA Zone XI Historic Preservation Commendation

2008 James B. Nutter

GCA Zone XI Horticulture Achievement Award

1977 Diana Hearne James (Mrs. Frederic)

GCA Zone XI Horticulture Commendation

2013 Alan Branhagen

GCA Zone XI Judging Achievement Award

2008 Norma Henry Sutherland (Mrs. Dwight Dierks)

GCA Zone XI Marilyn Sward Award for Photography

2010 Margaret Weatherly Hall (Mrs. Thomas Bryan III)
2018 Lyndon Gustin Chamberlain (Mrs. Richard Hunter)

GCA Club Awards

GCA Club Medal of Merit

1973 Barbara Forrester Rahm (Mrs. Phillip F.)

11.9 *Don't Touch Me Darling*

11.11 *Evening Tulip*

11.10 *Dusty Rose*

11.12 *Festival Floral*

1989 Lillian Marrs Diveley (Mrs. Rex L.)
1989 Julia Chandler Tinsman (Mrs. C. Humbert)
1994 Helen Delano Sutherland (Mrs. Herman Robert)
1999 Betty Clapp Robinson (Mrs. Kip)
2001 Marilyn Pierson Patterson (Mrs. Doyle)
2002 Norma Henry Sutherland (Mrs. Dwight Dierks)
2003 Hazel Hillix Barton Wells (Mrs. James W. Jr.)
2006 Suzanne "Susie" Slaughter Vawter
2007 Marie Bell Watson O'Hara (Mrs. Jack Butler)
2012 Blair Peppard Hyde (Mrs. Michael Theodore)
2015 Nancy Lee Smith Kemper (Mrs. Jonathan McBride)
2020 Margaret Weatherly Hall (Mrs. Thomas Bryan III)

GCA Club Appreciation Award

1995 Harriet "Hattie" Collins Byers (Mrs. Reed Phillip)
2000 Phoebe Hasek Bunting (Mrs. Henry Sharpe)
2006 Linda Baker Spencer (Mrs. Byron)
2007 Blair Peppard Hyde (Mrs. Michael Theodore)
2008 Dody Phinny Gates (Mrs. Lathrop Mead)
2009 Elizabeth "Betty" Kennedy Goodwin (Mrs. Frederick Merry Jr.)
2011 Ruth "Boots" Mathews Churchill (Mrs. James)
2013 Ann Parker North Readey (Mrs. Bartley John III)
2014 Marilyn Bartlett Hebenstreit (Mrs. James Bryant)
2014 Virginia "Ginny" Bedford McCanse (Mrs. Lynn)
2018 Ellen Zimmer Darling (Mrs. Albert Jr.)
2018 Alison Wiedeman Ward (Mrs. Scott Hardman)
2019 Laura Lee Carkener Grace
2021 Susan Ambler Spencer
2021 Laura Babcock Sutherland (Mrs. Todd Latta)
2023 Carolyn Steele Kroh (Mrs. George Piersol)

GCA Club Civic Improvement Award

2007 Lorelei Manning Gibson (Mrs. David William)

GCA Club Civic Improvement Commendation

2006 The City of Mission Hills
2010 Tulips on Troost—Durwin Rice
2013 Adelaide Cobb and Louis Larrick Ward
2014 Ben Sharda
2015 Jonathan McBride Kemper

2018 BoysGrow
2019 Trapp and Company

GCA Club Communications Award

2017 Sharon Wood Orr (Mrs. Richard Moffett)
2018 Laura Babcock Sutherland (Mrs. Todd Latta)

GCA Club Conservation Award

1992 Betty Clapp Robinson (Mrs. Kip)
1999 Martha "Mimi" Cate Newell (Mrs. Paul Dayton)
2007 Virginia "Gina" McDonald Miller
2017 Kathy Garrett Gates (Mrs. Kirkland Hayes)

GCA Club Conservation Commendation

2015 Kansas City WildLands/Jerry Smith Park

GCA Club Floral Design Achievement Award

1995 Marie Bell Watson O'Hara (Mrs. Jack Butler)
1999 Margaret Weatherly Hall (Mrs. Thomas Bryan III)
2002 Suzanne "Susie" Slaughter Vawter
2004 Carolyn Steele Kroh (Mrs. George Piersol)
2006 Barbara "Barbie" Rahm Reno (Mrs. Marion Jr.)
2008 Joy Laws Jones (Mrs. Russell Scarritt)
2010 DeSaix "DeeDee" Willson Adams
2016 Blair Peppard Hyde (Mrs. Michael Theodore)

GCA Club Historic Preservation Award

2001 Norma Henry Sutherland (Mrs. Dwight Dierks)

GCA Club Horticulture Award

1965 Mildred "Millie" Schwartzburg Hoover (Mrs. William Harold)
1975 Norma Henry Sutherland (Mrs. Dwight Dierks)
1980 Harriet "Hattie" Collins Byers (Mrs. Reed Phillip)
1993 Marie Bell Watson O'Hara (Mrs. Jack Butler)
1999 Kathy Garrett Gates (Mrs. Kirkland Hayes)
2003 Lorelei Manning Gibson (Mrs. David William)
2007 Betty Askren Kessinger (Mrs. William Buckner)
2009 Virginia "Ginny" Bedford McCanse (Mrs. Lynn)
2013 Carolyn Steele Kroh (Mrs. George Piersol)

11.13 *Evening Tulip*

11.15 *I Dare You*

11.14 *Have a Great Day*

11.16 *Lemony Rose*

GCA Club Horticulture Commendation

2018 The Kansas City Rose Society
2019 The Giving Grove
2021 George Piersol Kroh

GCA Club Judging Award

2024 Margaret Weatherly Hall (Mrs. Thomas Bryan III)

GCA Club Photography Award

2011 Elizabeth "Betty" Kennedy Goodwin (Mrs. Frederick Merry Jr.)
2017 Helen J. Jones Lea (Mrs. Albert Robert)
2022 Lyndon Gustin Chamberlain (Mrs. Richard Hunter)

Westport Garden Club Special Awards

WGC "Above and Beyond Award"

2003 Sally Ann Kemper Wood (Mrs. Thomas Jefferson Jr.)
2005 Sallie Bet Ridge Watson (Mrs. Raymond Etheridge Jr.)
2007 Virginia "Ginny" Bedford McCanse (Mrs. Lynn)
2008 Cynthia "Cindy" Rapelye Cowherd
2009 Alison Wiedeman Ward (Mrs. Scott Hardman)
2010 Margaret "Peggy" Kline Rooney (Mrs. Kevin Joseph)
2011 Lyndon Gustin Chamberlain (Mrs. Richard Hunter)
2012 Nancy Newell Stark (Mrs. Allan Breed)
2013 Catherine "Cacki" Price Smith (Mrs. Lawrence Earl IV)
2014 Suzanne "Susie" Slaughter Vawter
2015 Kathy Garrett Gates (Mrs. Kirkland Hayes)
2016 Marsha Fisher Moseley (Mrs. Richard Franklin)
2017 Eric Tschanz
2019 Jill Kathryn Stewart Bunting (Mrs. Olen George)
2020 Ann Bunting Milton (Mrs. Chad Earl)
2022 Susan Small Spaulding (Mrs. Charles Arthur III)
2023 Alison Bartlett Jager (Mrs. Eric Thor)
2024 Dody Phinny Gates Everist (Mrs. Brian Douglass)

WGC "The Deep-Rooted Perennial Award"

2014 Ellen Jurden Hockaday (Mrs. Irvine Oty Jr.)
2014 Eulalie "Eulie" Bartlett Zimmer (Mrs. Hugh Joseph)
2016 Norma Henry Sutherland (Mrs. Dwight Dierks)

2016 Suzanne "Susie" Slaughter Vawter
2017 Florence "Flip" Logan Kline (Mrs. William Patton)
2017 Sarah "Sally" Steele West (Mrs. Robert Hartley)
2018 Barbara "Barbie" Rahm Reno (Mrs. Marion Jr.)
2020 Elizabeth "Betty" Kennedy Goodwin (Mrs. Frederick Merry Jr.)
2022 Paget Gates Higgins (Mrs. Thomas Myron III)
2022 Wendy Jarman Powell (Mrs. George Everett III)
2024 Blair Peppard Hyde (Mrs. Michael Theodore)
2024 Sharon Wood Orr (Mrs. Richard Moffett)

The Norma Henry Sutherland Award

2008 Laura Lee Carkener Grace
2009 Laura Babcock Sutherland (Mrs. Todd Latta)
2010 Ellen Zimmer Darling (Mrs. Albert Jr.)
2012 Kristin Colt Goodwin (Mrs. John Kennedy)
2013 Susan Ambler Spencer
2014 Jo Meyer Missildine (Mrs. Gary Drue)
2015 Susan "Susie" Campbell Heddens (Mrs. Barret Spencer III)
2016 Pam Sutherland Gyllenborg (Mrs. John Eric)
2017 Rosalyn "Rozzie" Hargis Motter (Mrs. William Colhoun III)
2020 Mary Ann Huddleston Powell (Mrs. Nicholas Kuehn)
2022 Kristie Carlson Wolferman (Mrs. Frederick Ross)
2023 Diane Dennis Sloan (Mrs. Richard Morrison)
2024 Laura Woods Hammond (Mrs. Charles Edward)

11.17 *Parrot Tulip*

11.19 *Pink Beauty*

11.18 *Peacock Tulip*

11.20 *Ripples of Coral*

11.21 *Spring Green*

11.23 *Spring Racing Stripes*

11.22 *Spring Opening*

11.24 *Tango With Me*

11.25 *The Heart of the Matter*

11.27 *Wild Red*

11.26 *The Sunshine Girl*

11.28 *Yellow Rose of Texas*

Members of The Westport Garden Club

Active Members, 2025

DeSaix "DeeDee" Willson Adams (1985)

Elizabeth "Beth" Ritchie Alm (Mrs. Keith Lennart) (2019)

Catherine Paige Smith Bennish (Mrs. Christopher Nelson) (2024)

Ellen Stolzer Bolen (Mrs. Daniel Patrick) (2014)

Jill Kathryn Stewart Bunting (Mrs. Olen George) (1994)

Wendy Hockaday Burcham (Mrs. James Grant) (2008)

Lucy Wells Coulson (Mrs. Fred Nathan IV) (2019)

Cynthia "Cindy" Rapelye Cowherd (1999)

Janet Smith Curran (Mrs. D. Patrick) (2023)

Emily Bartlett Darling (2022)

Linda Evans (2014)

Dody Phinny Gates Everist (Mrs. Brian Douglass) (1990)

Kathy Garrett Gates (Mrs. Kirkland Hayes) (1990)

Lorelei Manning Gibson (Mrs. David William) (1994)

Kristin Colt Goodwin (Mrs. John Kennedy) (2009)

Jane "Janie" Hedrick Grant (Mrs. William Thomas II) (2023)

Pam Sutherland Gyllenborg (Mrs. John Eric) (2013)

Margaret Weatherly Hall (Mrs. Thomas Bryan III) (1988)

Laura Woods Hammond (Mrs. Charles Edward) (2021)

Allison Langstaff Harding (Mrs. John Darren) (2023)

Mary Levesque Harrison (Mrs. John Scott) (2013)

Susan "Susie" Campbell Heddens (Mrs. Barret Spencer III) (2011)

Paget Gates Higgins (Mrs. Thomas Myron III) (1989)

Julia Fields Jackson (Mrs. Benjamin Caleb) (2023)

Alison Bartlett Jager (Mrs. Eric Thor) (2001)

Jennifer Ball Jones (Mrs. Douglas E.) (2022)

Nancy Lee Smith Kemper (Mrs. Jonathan McBride) (1997)

Mary Campbell Kerr (2024)

Marianne Maurin Kilroy (Mrs. William Terrence) (2022)

Mary McPherson Kilroy (Mrs. John Muir Jr.) (2024)

Carolyn Steele Kroh (Mrs. George Piersol) (1986)

Ann Bunting Milton (Mrs. Chad Earl) (2010)

Jo Meyer Missildine (Mrs. Gary Drue) (2010)

Marsha Fisher Moseley (Mrs. Richard Franklin) (2010)

Rosalyn "Rozzie" Hargis Motter (Mrs. William Colhoun III) (2014)

Britton Franke Norden (Mrs. Richard G.) (2023)

Courtney Slaughter O'Farrell (Mrs. Patrick) (2023)

Sharon Wood Orr (Mrs. Richard Moffett) (1988)

Virginia "Ginger" McCord Owen (Mrs. Richard Fred) (2003)

Kristin Spicer Knight Patterson (Mrs. Evan David) (2023)

Susan Chadwick Pierson (Mrs. John T. Jr.) (2001)

Martha Mary Lally Platt (Mrs. Stephen Ernest) (2020)

Mary Ann Huddleston Powell (Mrs. Nicholas Kuehn) (2015)

Shelley Allen Preston (Mrs. David William) (2021)

Ann Parker North Readey (Mrs. Bartley John III) (1987)

Page Branton Reed (Mrs. Bruce Alan) (2023)

Margaret "Peggy" Kline Rooney (Mrs. Kevin Joseph) (2007)

Allie Ingram-Eiser Sifers (Mrs. Timothy Lyon) (2020)

Diane Dennis Sloan (Mrs. Richard Morrison) (2020)

Catherine "Cacki" Price Smith (Mrs. Lawrence Earl IV) (1995)

Melinda "Mindy" Scudder Sosland (Mrs. Meyer Joseph) (2024)

Susan Small Spaulding (Mrs. Charles Arthur III) (2013)

Susan Ambler Spencer (2011)

Laura Babcock Sutherland (Mrs. Todd Latta) (2007)

Nancy Booth Valentine (Mrs. John West Jr.) (2023)

Alison Wiedeman Ward (Mrs. Scott Hardman) (2003)

Marcy Dickey Waugh (Mrs. George Lapsley) (2023)

Charlotte Russell White (Mrs. Dwayne Kent) (2022)

Kristie Carlson Wolferman (Mrs. Frederick Ross) (2019)

Honorary Members

Robert J. "Bob" Berkebile (Libby) (2009)

Benjamin "Ben" Sharda (Laurie) (2015)

Eric Tschanz (Debbie) (2006)

Affiliate Members

Kimberly "Kim" Kline Aliber (Mrs. William John) (2006)

Newell Gates Brookfield (Mrs. Charles Cameron) (2020)

Lyndon Gustin Chamberlain (Mrs. Richard Hunter) (2004)

Elizabeth "Betty" Kennedy Goodwin (Mrs. Frederick Merry Jr.) (1982)

Laura Lee Carkener Grace (2002)

Carlene Barrett Hall (Mrs. William Austin) (2004)

Marilyn Bartlett Hebenstreit (Mrs. James Bryant) (2003)

Ellen Jurden Hockaday (Mrs. Irvine Oty Jr.) (1987)

Blair Peppard Hyde (Mrs. Michael Theodore) (2000)

Ellie Aliabadi Idstrom (Mrs. Mark E.) (2010)

Betty Askren Kessinger (Mrs. William Buckner) (1993)

Diana Jackson Shand Kline (Mrs. Leonard Patton Jr.) (2008)

Florence "Flip" Logan Kline (Mrs. William Patton) (1975)

Helen J. Jones Lea (Mrs. Albert Robert) (2002)

Virginia "Ginny" Bedford McCanse (Mrs. Lynn) (2004)

Virginia "Gina" McDonald Miller (1984)

Catherine "Kay" Strick Newell (Mrs. Charles Mason) (2014)

Marie Bell Watson O'Hara (Mrs. Jack Butler) (1970)

Ellen Kirwan Porter (Mrs. Graham) (2003)

Laura Keller Powell (Mrs. Richardson Kammeyer) (2010)

Wendy Jarman Powell (Mrs. George Everett III) (1996)

Jan Dye Riss (Mrs. Edward Stayton) (1992)

Katherine "Kathy" Ehlers Sawyer (Mrs. Joseph Neal) (1997)

Nancy Newell Stark (Mrs. Allan Breed) (2001)

Megan Mahaffy Sutherland (Mrs. Mark Breidenthal) (2015)

Nancy Embry Thiessen (Mrs. Michael Robert) (2007)

Barbara Farmer Thompson (Mrs. Webster Townley) (2009)

Sarah "Sally" Steele West (Mrs. Robert Hartley) (1992)

Presidents of The Westport Garden Club

Founding Member of The Westport Garden Club

1950–1951 *Mildred "Millie" Schwartzburg Hoover (Mrs. William Harold)

1951–1952 *Katherine "Katie" Kessinger (Mrs. Joseph Webb)

1952–1953 *Helen Hayes Thompson (Mrs. Mason Leo)

1953–1954 *Maxine Maxwell Goodwin (Mrs. Frederick Merry)

1954–1955 *Lillian Marrs Diveley (Mrs. Rex L.)

1955–1956 *Laura Kemper Toll Carkener (Mrs. George Guyton)

1956–1957 *Helen Elizabeth "H.E." McCune Hockaday (Mrs. Irvine Oty)

1957–1958 Virginia Page Hart (Mrs. Creighton Carlton)

1958–1959 *Julia Chandler Tinsman (Mrs. Charles Humbert)

1959–1961 *Josephine "Jo" Reid Stubbs (Mrs. Charles Stephen III)

1961–1963 Vivian Pew Foster (Mrs. Francis Gregg)

1963–1965 Martha Callaway Knight (Mrs. John Swann)

1965–1967 *Laura Kemper Toll Carkener (Mrs. George Guyton)

1967–1969 Helen Delano Sutherland (Mrs. Herman Robert)

1969–1971 Elizabeth "Libby" Scarritt Adams (Mrs. Benjamin Cullen Jr.)

1971–1973 Katherine "Kitty" Hall Wagstaff (Mrs. Robert Wilson)

1973–1975 Jean Holmes McDonald (Mrs. William M.)

1975–1977 Diana Hearne James (Mrs. Frederic)

1977–1979 Norma Henry Sutherland (Mrs. Dwight Dierks)

1979–1981 Barbara Shackelford Seidlitz (Mrs. Charles Newman Jr.)

1981–1983 Sarah "Sally" Rapelye Cowherd (Mrs. Grant)

1983–1985 Sally Ann Kemper Wood (Mrs. Thomas Jefferson Jr.)

1985–1987 Marie Bell Watson O'Hara (Mrs. Jack Butler)

1987–1989 Sallie Bet Ridge Watson (Mrs. Raymond Etheridge Jr.)

1989–1991 Suzanne "Susie" Slaughter Vawter

1991–1993 Marilyn Pierson Patterson (Mrs. John Doyle)

1993–1995 Hazel Hillix Barton (Mrs. Clarence Dickinson)

1995–1997 Margaret Weatherly Hall (Mrs. Thomas Bryan III)

1997–1999 Elizabeth "Betty" Kennedy Goodwin (Mrs. Frederick Merry Jr.)

1999–2001 Eulalie "Eulie" Bartlett Zimmer (Mrs. Hugh Joseph Jr.)

2001–2003 Ellen Jurden Hockaday (Mrs. Irvine Oty Jr.)

2003–2005 Paget Gates Higgins (Mrs. Thomas Myron III)

2005–2007 Sarah "Sally" Steele West (Mrs. Robert Hartley)

2007–2009 Ann Parker North Readey (Mrs. Bartley John III)

2009–2011 Blair Peppard Hyde (Mrs. Michael Theodore)

2011–2013 Betty Askren Kessinger (Mrs. William Buckner)

2013–2015 Nancy Lee Smith Kemper (Mrs. Jonathan McBride)

2015–2017 Laura Babcock Sutherland (Mrs. Todd Latta)

2017–2019 Cynthia "Cindy" Rapelye Cowherd

2019–2021 Carolyn Steele Kroh (Mrs. George Piersol)

2021–2023 Margaret "Peggy" Kline Rooney (Mrs. Kevin Joseph)

2023–2025 Mary Ann Huddleston Powell (Mrs. Nicholas Kuehn)

Members of The Westport Garden Club Who Served in Garden Club of America Positions

Elizabeth Scarritt Adams
Zone XI Chair, 1971–1973
Admissions Committee, Zone Rep, 1973–1975
Admissions Committee, Chair, 1975–1977
Interchange Fellowship Committee, Zone Rep, 1977–1978
Director, 1978–1980
Board of Associates, 1977

Hazel Hillix Barton
Visiting Gardens Committee, Zone Rep, 1997–1999

Florence Boyd Beaham
Floral Design Judge, 1983

Josephine Jobes Giles Bunting
International Flower Show Committee, Member, 1964–1965

Lyndon Gustin Chamberlain
Photography Judge, 2019
Photography Committee, Assistant Coordinator, Photography Study Group, 2020
Photography Committee, Coordinator, Photography Study Group, 2020–2022

Lillian Marrs Diveley
Floral Design Judge, 1983

Elizabeth Kennedy Goodwin
Garden History and Design Committee, Zone Rep, 1999–2001
Visiting Gardens Committee, Zone Rep, 2001–2003
Annual Meeting, Co-Chair, 2004–2005
Public Relations Committee, Zone Rep, 2005–2006
Garden History and Design Committee, Chair, 2006–2008
Horticulture Judge, 2010
Board of Associates, 2005

Margaret Weatherly Hall
Floral Design Judge, 1995
Zone Chair, 1999–2001
Annual Meeting, Co-Chair, 2004–2005
Nominating Committee, Zone Rep, 2005–2007
Admissions Committee, Chair, 2007–2009
Photography Judge, 2007
Director, 2009–2011
Director Serving on the Executive Committee, 2010–2011
Corresponding Secretary, 2011–2013
Stewardship Committee, Member, 2019–2022
Board of Associates, 2001

Virginia Page Hart
International Flower Show Committee, Member, 1964

Paget Gates Higgins
Public Relations Committee, Zone Rep, 2006–2007

Mildred Schwartzburg Hoover
Horticulture Committee, Zone Rep, 1965–1968

Diana Hearne James
Horticulture Committee, Vice Chair, 1977–1979

Nancy Lee Smith Kemper
Founders Fund Committee, Zone Rep, 2015–2017
Visiting Gardens Committee, Zone Rep, 2017–2019

Ruth Carr Patton Kline
Director, 1961–1964
Medal Awards Committee, Member, 1965

Carolyn Steele Kroh
Visiting Gardens Committee, Zone Rep, 2021–2023

Virginia McDonald Miller
Conservation Committee, Zone Rep, 2013–2015
National Affairs and Legislation Committee, Zone Rep, 2013–2015

Helen Houston Nelson
Headquarters Decorating Committee, Member, 1964

Betty Clapp Robinson
Conservation Committee, Zone Rep, 1965–1967
Conservation Committee, Vice Chair, Environmental Quality, 1975–1977
Legislative Committee, Member, 1975–1977
Conservation Committee, Chair, 1977–1979
Nominating Committee, Zone Rep, 1979–1981
Nominating Committee, Chair, 1980
Admissions Committee, Chair, 1981–1983
Bulletin Committee, Zone Rep, 1984

Director, 1985–1987
Slide Library, Member, 1988–1995
Archives of American Gardens, Member, 1995
Board of Associates, 1979

Maxine Christopher Shutz
International Flower Show Committee, Member, 1966

Josephine Reid Stubbs
Public Relations Committee, Zone Rep, 1963–1965
Program Committee, Zone Rep, 1973–1975
The American Academy in Rome, GCA Liaison and Rome Prize Selection Committee, Member, 1986–1994

Helen Delano Sutherland
Bulletin Committee, Zone Rep, 1969–1972

Laura Babcock Sutherland
Awards Committee, Zone Rep, 2017–2019
Communications Committee, Zone Rep, 2017–2019
Communications Committee, Vice Chair, Public Pages, 2020–2022
Communications Committee, First Vice Chair, 2022–2023

Norma Henry Sutherland
Horticulture Committee, Zone Rep, 1973–1975
Interchange Fellowship, Selection Committee, Member, 1981
Awards Committee, Zone Rep, 1981–1983
Zone Chair, 1983–1985
Flower Design Judge, 1984
Judging Committee, Zone Rep, 1985–1987
Horticulture Judge, 1986
Director, 1990–1992
Vice President, 1993–1995
Flower Show Committee, Area Vice Chair, 1996–1997
Endowment Committee, Member, 2002–2005
Annual Meeting, Honorary Chair, 2005
Board of Associates, 1985

Julia Chandler Tinsman
Judging Committee, Zone Rep, 1963–1965
Judging Committee, Vice Chair, 1973–1975
Director, 1975–1977
Awards Committee, Zone Rep, 1977–1979
Judging Committee, Vice Chair, 1983–1985
Flower Arranging Judge, 1983
Board of Associates, 1977

Suzanne Slaughter Vawter
Zone Vice Chair, 1991–1993
Scholarship Committee, Zone Rep, 1994–1995

In Memoriam

*Ulva "Jimmy" Narr (Mrs. Frederick)

*Ruth Rubey Kemper (Mrs. James Madison)

*Genevieve Marcell Davis (Mrs. Manvel H.)

*Eleanor Nichols "Nicky" Allen (Mrs. Earl Wilson)

*Maude Chatten Cowherd (Mrs. Fletcher Jr.)

*Jane Hemingway Gordon (Mrs. George L.)

*Elizabeth "Liz" Nesbit Marty (Mrs. Samuel C.)

*Katherine "Katy" Fisher Satterlee (Mrs. William Bertrand)

*Sarah Bryant Merriman (Mrs. Jack Dawson)

*Airy Smeltzer Jones (Mrs. Cliff C.)

*Leila Grant Cowherd (Mrs. Joseph B.)

*Helen Hayes Thompson (Mrs. Mason Leo)

Charlotte Carnes Kitchen (Mrs. Lewis)

Clarence Halsell Holmes (Mrs. Jay V.)

*Mildred "Millie" Schwartzburg Hoover (Mrs. William Harold)

*Enid Jackson Kemper (Mrs. Rufus Crosby)

*Katherine "Katie" Buckner Kessinger (Mrs. Joseph Webb)

*Mary Dickinson Barton (Mrs. George Allen Jr.)

Helen Foresman Spencer (Mrs. Kenneth Aldred)

*Ann Peppard White (Mrs. James Mayne)

*Mildred Peet Welch (Mrs. Gustavus D.)

Sarah "Sally" Rapelye Cowherd (Mrs. Grant)

*Patti Harding Abernathy (Mrs. Taylor Stevenson)

*Kathleen I. Roth (Mrs. Richard G.)

Martha Callaway Knight (Mrs. John Swann)

*Helen Ward Beals (Mrs. David Thomas Jr.)

*Flora Markey Barton (Mrs. Francis Waddell)

*Maxine Christopher Shutz (Mrs. Byron Theodore)

*Marcella Ryan Peppard (Mrs. Joseph Greer)

*Barbara Forrester Rahm (Mrs. Phillip F.)

*Pauline Atterbury Dierks (Mrs. DeVere)

Georgette Longan O'Brien (Mrs. Terence M.)

Mildred "Millie" Lane Kemper (Mrs. James Madison Jr.)

*Marjorie Lane Paxton (Mrs. Frank)

*Dorothy Dillon Cunningham (Mrs. John R.)

Virginia Sartor Foresman (Mrs. Max W.)

*Lillian Marrs Diveley (Mrs. Rex L.)

*Helen Elizabeth "H.E." McCune Hockaday (Mrs. Irvine Oty)

*Patricia "Patty" Castle White (Mrs. Raymond B.)

*Florence Boyd Beaham (Mrs. Gordon T. Jr.)

*Newell "Honey Boy" McGee Townley (Mrs. Webster W.) Thornton (Mrs. Oliver Comstock)

*Hester "Hettie" Milligan Gustin (Mrs. Albert L. Jr.)

Marion Enggas Kreamer (Mrs. John Harrison)

*Mary Histed Hughes (Mrs. Hilliard Withers)

*Ruth Carr Patton Kline (Mrs. Leonard Charles)

*Julia Jackman Bartlett (Mrs. Francis Wayland Jr.)

Madeline Sweeney Smith (Mrs. Lawrence Earl Jr.)

Elizabeth "Libby" Scarritt Adams (Mrs. Benjamin Cullen Jr.)

*Maxine Maxwell Goodwin (Mrs. Frederick Merry)

*Marguerite "Margot" Munger Peet (Mrs. Herbert O.)

*Eda Marie Peck Luger (Mrs. Charles Russell)

*Katherine Histed Foster (Mrs. William H.)

*Barbara James McGreevy (Mrs. Milton W.)

Vivian Pew Foster (Mrs. Francis Gregg)

Diana Hearne James (Mrs. Frederic)

Betty Berry McLaughlin (Mrs. Gene R.)

Virginia "Ginny" Page Hart (Mrs. Creighton Carlton)

Mary Louise "Mary Lou" Harris Blackwell (Mrs. Menefee D.)

*Ruth Flower Lester (Mrs. Robert R.)

*Thelma Williams Frick (Mrs. Fred C.)

Margaret "Peggy Sue" Neal

Betty Clapp Robinson (Mrs. Kip)

Josephine "Josie" Jobes Giles Bunting (Mrs. Clarke S. P.)

*Julia Chandler Tinsman (Mrs. Charles Humbert)

Virginia Haynes Weatherly

Doris "Dorry" Mead Gates (Mrs. Clinton)

Virginia Wilber Torrance Bolin (Mrs. Frank E.)

Helen Houston Nelson (Mrs. Richard Robinson)

Katherine "Kitty" Hall Wagstaff (Mrs. Robert Wilson)

Helen Delano Sutherland (Mrs. Herman Robert)

Elizabeth "Betty" Searle Thompson (Mrs. Mason Leo Jr.)

Alice "Allie" Parker Scarritt Kelley (Mrs. John Colt)

Sallie Bet Ridge Watson (Mrs. Raymond Etheridge Jr.)

Harriet "Hattie" Collins Byers (Mrs. Reed Phillip)

*Virginia French Mackie (Mrs. David Charles)

*Josephine "Jo" Reid Stubbs (Mrs. Charles Stephen III)

Dorothy "Dee" Halsey Hughes (Mrs. David Histed)

Linda Baker Spencer (Mrs. Byron Jr.)

Barbara Welch Thompson (Mrs. Hoyt Hayes)

Jeanne McCray Beals (Mrs. David Thomas III)

*Sally Sheffield Ingalls Keith (Mrs. Edward)

*Laura Kemper Toll Carkener (Mrs. George Guyton)

Wendy Byers Brinton (Mrs. William R.)

Martha "Mimi" Cate Newell (Mrs. Paul Dayton)

Jean Holmes McDonald Deacy (Mrs. Thomas E.)

Sally Ann Kemper Wood (Mrs. Thomas Jefferson Jr.)

Ellison "Kelly" Brent Lambert (Mrs. Sanders Ray Jr.)

Marilyn Pierson Patterson (Mrs. John Doyle)

Mariel Tyler Thompson (Mrs. Edwin H. Leo II)

Adele Coryell Hall (Mrs. Donald Joyce)

Laura Lane Kemper Fields (Mrs. Michael Duane)

Hazel Hillix Barton (Mrs. Clarence Dickinson)

Laura Ann Shutz Cray (Mrs. Richard Baxter)

Margaret "Peggy" Garner Gustin (Mrs. Albert Lyman III)

Prudence "Prue" Townley Thompson (Mrs. Hoyt Hayes)

Barbara Shackelford Seidlitz (Mrs. Charles Newman Jr.)

Joan "Jody" Jenkins Bartlett (Mrs. Paul Dana Jr.)

Barbara "Barbie" Rahm Reno (Mrs. Marion Jr.)

Karen Iverson Bartlett (Mrs. D. Brook)

Ellen Zimmer Darling (Mrs. Albert Jr.)

Eulalie "Eulie" Bartlett Zimmer (Mrs. Hugh Joseph)

Norma Henry Sutherland (Mrs. Dwight Dierks)

Jeanne Forney Bleakley (Mrs. Charles E.)

Suzanne "Susie" Slaughter Vawter

Joy Laws Jones (Mrs. Russell Scarritt)

Phoebe Hasek Bunting (Mrs. Henry Sharpe)

Notes

Chapter 1
The Westport Garden Club Sprouts

1. "Memories of Miss Hull," May 2023, Ridgefield Garden Club Archives; "Col. Abernathy Dead," *Lawrence Weekly World*, December 18, 1902.

2. Linda Mack, *Madeline Island Summer Houses: An Intimate Journey*, 69; society page, *Kansas City Star*, September 28, 1930; Betsy Cullen, "Elizabeth Abernathy Hull," 29–31.

3. Candid photo of Mrs. Mason L. Thompson, with cutline, *The Independent*, July 1, 1950, 3.

4. "2023 Memories of Miss Hull."

5. "Elizabeth Abernathy Hull Award," The Garden Club of America, https://www.gcamerica.org/members/civic-improvement-manual, accessed June 25, 2020.

6. Mildred S. Hoover, report, "How It All Began," 1, *The Westport Garden Club's Twenty-fifth Anniversary Program*, History Notebook, WGC Archives (WGCA).

7. Hoover, "How It All Began," 1.

8. Hoover, "How It All Began," 1.

9. "About Linda Hall Library," Linda Hall Library, https://www.lindahall.org/about/our-story, accessed May 2, 2020.

10. Lillian Diveley, report, "History of The Westport Garden Club," 13, *The Westport Garden Club's Twenty-fifth Anniversary Program*, History Notebook, WGCA.

11. "Introduction to the Linda Hall Library Arboretum," Linda Hall Library, https://www.lindahall.org/arboretum; "KC Metro Champion Trees," Bridging the Gap, https://www.bridgingthegap.org/kc-metro-champion-trees.

12. Diveley, "History of The Westport Garden Club," 2.

13. Vivian Foster, report, "Civic Projects," 2, *The Westport Garden Club's Twenty-fifth Anniversary Program*, History Notebook, WGCA.

14. Foster, "Civic Projects," 2.

15. "Flood of 1951," *Kansapedia*, Kansas Historical Society, https://www.kshs.org/kansapedia/flood-of-1951/17163, accessed May 5, 2020.

16. Monroe Dodd, "How Floods Shaped the Kansas City We Know Today."

17. "Westport Garden Club at Sleepy Hill Farm," *The Independent*, July 21, 1951.

18. Diveley, "History of The Westport Garden Club," 3.

19. "Control on Tree Men," *Kansas City Star*, December 31, 1951, clipping in Westport Garden Club Scrapbook: 1950–1958, WGCA.

20. Diveley, "History of The Westport Garden Club," 3, 4.

21. "Busily at Work on Their Major Landscaping Effort of the Season," *Kansas City Star*, April 1, 1952, clipping in Westport Garden Club Scrapbook: 1950–1958, WGCA.

22. "Project No. 1," *The Independent*, April 4, 1952; "Busily at Work."

23. Foster, "Civic Projects," 2.

24. Lewis Kitchen to Katie Kessinger, February 13, 1952, Westport Garden Club Scrapbook: 1950–1958, WGCA.

25. Foster, "Civic Projects," 1.

26. Gordon and Gilmore to Maxine M. Goodwin, September 21, 1953, articles of incorporation folder, WGCA.

27. "State of Missouri Certificate of Amendment," March 18, 1963; Lathrop, Righter, Gordon, and Parker to Mrs. Francis G. [Vivian] Foster, May 6, 1963; Lawrence B. Jerome to The Westport Garden Club, October 20, 1965, all in articles of incorporation folder, WGCA.

28. Foster, "Civic Projects," 3.

29. William Rockhill Nelson Trust, Trustees' Minutes, December 2, 1954, vol. 9, Nelson-Atkins Museum of Art Archives (NAMAA).

30. Katie Kessinger, report, "Festival Report," 1, *The Westport Garden Club's Twenty-fifth Anniversary Program*; *Westport Garden Club Flower Festival*, brochure, both in History Notebook, WGCA; "Flower Festival at Gallery," *The Independent*, May 28, 1955.

31. Kessinger, "Festival Report," 1.

32. Kessinger, "Festival Report," 1; Laurence Sickman, "Comments of the Director," *Annual Report 1955*, NAMAA; Nelson Trust, Trustees' Minutes, July 14, 1955, vol. 10, NAMAA.

33. Kessinger, "Festival Report," 2.

34. "Flower Festival," *The Independent*, May 26, 1956.

35. "Flower Festival," *The Independent*; Sickman, "Comments of the Director," 1, and "Special Events," 79, both in *Annual Report 1956*, NAMAA.

Chapter 2
The Garden Club of America: Strength in Numbers

1. "About Us," The Garden Club of Philadelphia, https://www.thegardenclubofphiladelphia.org/about, accessed June 6, 2020.

2. Ernestine Abercrombie Goodman, *The Garden Club of America: History 1913–1938*, 7; "About Us," The Garden Club of Philadelphia.

3. Goodman, foreword to *The Garden Club of America*.

4. Goodman, *The Garden Club of America*, 8; Marjorie Gibbon Battles and Catherine Colt Dickey, *Fifty Blooming Years, 1913–1963*, 17.

5. *Bulletin*, no. 1, 1913, 1; Kate Brewster, *Bulletin*, no. V (second series), July 1920, 2. All *Bulletin* articles cited are accessible to members on the GCA website.

6. *Bulletin*, no. IV, April 1914, 8.

7. Battles and Dickey, *Fifty Blooming Years*, 19.

8. Battles and Dickey, *Fifty Blooming Years*, 19; William Seale, *The Garden Club of America: A Hundred Years of a Growing Legacy*, 11–13.

9. Elizabeth Martin, "The President's Appeal," *Bulletin*, no. XXI, July 1917.

10. "Annual Meeting," *Bulletin*, no. XXV, May 1918.

11. Battles and Dickey, *Fifty Blooming Years*, 29.

12. Seale, *The Garden Club of America*, 69.

13. Battles and Dickey, *Fifty Blooming Years*, 31; "Flower Shows," The Garden Club of America, https://www.gcamerica.org/what-we-do-flower-shows, accessed June 29, 2020.

14. "National Gardens," *Bulletin*, no. III, January 1914.

15. "National Gardens," *Bulletin*, no. IV, April 1914.

16. Battles and Dickey, *Fifty Blooming Years*, 63–64.

17. Battles and Dickey, *Fifty Blooming Years*, 63–64; Seale, *The Garden Club of America*, 58.

18. Battles and Dickey, *Fifty Blooming Years*, 29; Claire P. Caudill, Betty Pinkerton, Diane B. Stoner, and Dede Petri, *A History of Conservation and National Affairs and Legislation: The Garden Club of America, 1913–2019*, 8; Seale, *The Garden Club of America*, 42–45.

19. Seale, *The Garden Club of America*, 104.

20. *Bulletin*, no. V, July 1914; Seale, *The Garden Club of America*, 22–23.

21. Caudill, Pinkerton, Stoner, and Petri, *History of Conservation*, 8; Battles and Dickey, *Fifty Blooming Years*, 58–59.

22. Caudill, Pinkerton, Stoner, and Petri, *History of Conservation*, 8.

23. Caudill, Pinkerton, Stoner, and Petri, *History of Conservation*, 9–10; Seale, *The Garden Club of America*, 50–53; Battles and Dickey, *Fifty Blooming Years*, 61.

24. Caudill, Pinkerton, Stoner, and Petri, *History of Conservation*, 9–10, 11; Seale, *The Garden Club of America*, 53; Battles and Dickey, *Fifty Blooming Years*, 61, 71.

25. Seale, *The Garden Club of America*, 53; "The GCA and Save the Redwoods League," The Garden Club of America, March 19, 2019, https://www.gcamerica.org/news/get/id/2376.

26. Caudill, Pinkerton, Stoner, and Petri, *History of Conservation*, 9–10, 11; "The GCA and Save the Redwoods League."

27. Caudill, Pinkerton, Stoner, and Petri, *History of Conservation*, 9–10, 11; Janet Riley, "GCA Medalist in the News: Save the Redwoods League Acquired Lost Coast Redwoods Property," The Garden Club of America, March 11, 2022, https://www.gcamerica.org/news/get?id=3744.

28. "Redwoods' Ability to Store Carbon Determined Invaluable," The Garden Club of America, July 14, 2020, https://www.gcamerica.org/news/get/id/2961.

29. Seale, *The Garden Club of America*, 80–81.

30. Seale, *The Garden Club of America*, 83–89.

31. Seale, *The Garden Club of America*, 79, 91–92.

32. Caudill, Pinkerton, Stoner, and Petri, *History of Conservation*, 12.

33. Caudill, Pinkerton, Stoner, and Petri, *History of Conservation*, 12.

34. Caudill, Pinkerton, Stoner, and Petri, *History of Conservation*, 12.

35. Battles and Dickey, *Fifty Blooming Years*, 70–71.

Chapter 3
The Early Years, 1956–1969

1. Card written by Helen Hockaday; Lillian Diveley, report, "History of the Westport Garden Club," *The Westport Garden Club's Twenty-fifth Anniversary Program*, both in History Notebook, WGCA.

2. Diveley, "History of the Westport Garden Club," 5.

3. "Third Annual Flower Festival," *The Independent*, April 6, 1957; Katie Kessinger, report, "Festival Report," 2, *The Westport Garden Club's Twenty-fifth Anniversary Program*, History Notebook, WGCA; Laurence Sickman, "Comments of the Director," *Annual Report 1957*, NAMAA.

4. "Busy Days for Club," *The Independent*, September 21, 1957.

5. "Zone Meeting," *The Independent*, October 12, 1957; Mrs. Harold H. Seaman, "Central Western Zone Meeting," Westport Garden Club Scrapbook: 1950–1958, WGCA.

6. Seaman, "Central Western Zone Meeting."

7. Debbie Thompson Gates to author, email, January 7, 2020; "Nelson Gallery to Host 4,000," *The Independent*, June 13, 1960; "Primitive Art Show Opening Is Social Event," *Kansas City Star*, January 21, 1962, clipping in Westport Garden Club Scrapbook: 1961–1966; "Turkish Exhibit," image in Westport Garden Club Scrapbook: 1967–1970, last two in WGCA.

8. Diveley, "History of the Westport Garden Club," 9–10.

9. "Floral Table Arrangements," Westport Garden Club Scrapbook: 1981–1985, WGCA.

10. "Mrs. Frank Paxton," Westport Garden Club Scrapbook: 1961–1966, WGCA; "The Ivy Eagle," *The Independent*, June 16, 1962; Bruce Barton to author, email, January 27, 2022.

11. "Beauty Keynotes Luncheon," *The Independent*, December 17, 1960; "Legends of Christmas," *The Independent*, December 12, 1987.

12. "Legends of Christmas," *The Independent*; DeSaix Adams, interview by author, January 3, 2022.

13. Laura Kemper Carkener, "The Plants of Christmas"; "Traditions of Christmas, 2012," foreword by Betty Kennedy Goodwin, both in History Notebook, WGCA.

14. Kessinger, "Festival Report," 2.

15. William Rockhill Nelson Trust, Trustees' Minutes, September 30, 1958, vol. X, 1955–1958, NAMAA; "Flower Festivals," box 1, folder 2, Westport Garden Club Records, RG 71/01, NAMAA; "Fashions for Garden Parties and Garden Digging," *Kansas City Star*, April 23, 1959, clipping in Westport Garden Club Scrapbook: 1958–1961, WGCA.

16. Kessinger, "Festival Report," 4; "Westport Garden Club Flower Festival 1966: Chairman's Report," WGCA.

17. "Junior Gallery and Creative Arts Center," box 1, folder 1, Westport Garden Club Records, RG 71/01, NAMAA.

18. Ann Brubaker, "History of the Education Department," November 1980, Department of Education Records, box 1, RG 32, NAMAA; *Choice Cuts*, published by the Junior League of Kansas City, Missouri, vol. XXXII, no. 5, January 1960, box 1, folder 1, Westport Garden Club Records, RG 71/01, NAMAA.

19. Kessinger, "Festival Report," 4.

20. Invitation to Junior Gallery opening, box 1, folder 1, Westport Garden Club Records, RG 71/01, NAMAA.

21. "About 300 at Opening at New Junior Gallery," *Kansas City Star*, January 30, 1960, clipping in box 1, folder 1, Westport Garden Club Records, RG 71/01, NAMAA.

22. "About 300 at Opening"; Louise Conduit, Trustees' invitation, Westport Garden Club Scrapbook: 1958–1961, WGCA.

23. "Junior Gallery," Department of Education Records, RG 32/01, NAMAA; Jan Dickerson, "The Wonders of the World of Pictures," *Kansas City Star*, clipping in box 1, folder 1, Westport Garden Club Records, RG 71/01, NAMAA; "Report: The Director," box 1, folder 1, Westport Garden Club Records, RG 71/01, NAMAA.

24. Kessinger, "Festival Report," 4.

25. "Katherine Harvey," Westport Garden Club Scrapbook: 1958–1961, WGCA; Blair Peppard Hyde, interview by author, June 8, 2022.

26. "Report: The Director," box 1, folder 1, Westport Garden Club Records, RG 71/01; "Medal to Garden Club for Junior Art Gallery," *Kansas City Times*, May 21, 1964, clipping in box 1, folder 2, Westport Garden Club Records, RG 71/01, both in NAMAA.

27. James Seidelman, "A Junior Gallery Within a Large Museum," *Museum News*, April 1961, clipping in Westport Garden Club Scrapbook: 1958–1961, WGCA.

28. Seidelman, "A Junior Gallery."

29. "Club Season Ends Brilliantly," *The Independent*, June 17, 1961.

30. "Garden Tour by Westport Garden Club," *The Independent*, June 16, 1962.

31. Kessinger, "Festival Report," 5.

32. "Flower Festival," *The Independent*, February 4, 1961; Kessinger, "Festival Report," 5.

33. Kessinger, "Festival Report," 5.

34. "Preparations for Sixth Flower Festival," *The Independent*, April 20, 1963; "Flower Festivals," box 1, folder 1, Westport Garden Club Records, RG 71/01, NAMAA.

35. "Rozzelle Court of Nelson Gallery," *The Independent*, July 2, 1966; Jean Holmes McDonald, "Westport Garden Club Flower Festival 1966: Chairman's Report"; Larry Sickman to Jean McDonald, May 19, 1966, last two in WGCA.

36. "Collector's Corner," Westport Garden Club Scrapbook: 1967–1970, WGCA; "A Tradition Continues at the Nelson Gallery," *The Independent*, April 5, 1969.

37. Kessinger, "Festival Report," 8.

38. Mrs. Charles N. [Barbara] Seidlitz Jr. to Marc Wilson, March 26, 1985; "Westport Garden Club Flower Festival 1985: Chairman's Report," both in Westport Garden Club Scrapbook: 1981–1985, WGCA.

39. Laurence Sickman, "Comments of the Director," *Annual Report 1969*, NAMAA.

40. Marc Wilson to author, email, April 10, 2024.

41. Kessinger, "Festival Report," 9; "GCA Statement of Purpose," Garden Club of America, https://www.gcamerica.org/about, accessed November 7, 2024.

Chapter 4
The Westport Garden Club and the
Environmental Movement in the 1960s and 1970s

1. Claire P. Caudill, Betty Pinkerton, Diane B. Stoner, and Dede Petri, *A History of Conservation and National Affairs and Legislation: The Garden Club of America, 1913–2019*, 12.

2. "The Lakeside Nature Center," *The Independent*, May 2, 1970.

3. Caudill, Pinkerton, Stoner, and Petri, *History of Conservation*, 17.

4. Rachel Carson, *Silent Spring*, 97; Ann H. Convery, GCA Library Chairman, to Mrs. Joseph Wittman, March 7, 1997, Ridgefield Garden Club Archives.

5. Caudill, Pinkerton, Stoner, and Petri, *History of Conservation*, 14–15.

6. Caudill, Pinkerton, Stoner, and Petri, *History of Conservation*, 14–15; "Litterbug Campaign," WGC notes, June 1957, History Notebook, WGCA.

7. Caudill, Pinkerton, Stoner, and Petri, *History of Conservation*, 13.

8. WGC Membership Meeting Minutes, June 7, 2006, minutes 2005–2006 folder,

in Sally West, president's file, WGCA; Biography.com Editors and Tyler Piccotti, "Lady Bird Johnson."

9. "Scenic America Remembers Marion Fuller Brown," Scenic America, September 18, 2019, https://www.scenic.org/2019/09/18/scenic-america-remembers-marion-fuller-brown/.

10. Caudill, Pinkerton, Stoner, and Petri, *History of Conservation*, 20; Katherine Wagstaff, *President's Annual Report, June 1973*, WGCA.

11. Caudill, Pinkerton, Stoner, and Petri, *History of Conservation*, 16.

12. Caudill, Pinkerton, Stoner, and Petri, *History of Conservation*, 16.

13. WGC Board Meeting Minutes, May 28, 1969, Helen Sutherland, president's file, WGCA.

14. "Founders Fund Project, 1971," *Bulletin*, in Westport Garden Club Scrapbook: 1970–1973, WGCA.

15. Katie Kessinger, report, "Festival Report," 9, *The Westport Garden Club's Twenty-fifth Anniversary Program*, History Notebook, WGCA.

16. *The Garden Club of America Zone XI Meeting, April 20th, 21st, 22nd, 1971, Kansas City, Missouri*, booklet, in WGCA.

17. Lillian Diveley, "The Westport Garden Club Annual Meeting," June 2, 1971, Elizabeth Adams, president's file, WGCA.

18. *Symposium on the Ecology of Midwest Flyways and Prairie Flora*, booklet, in symposium notebook, WGCA. The following discussion is also from this booklet.

19. "Westport Garden Club Gives $200 to Northwest," *The Johnson County Sun*, May 31, 1974, clipping in flower festivals folder, WGCA.

20. Tom Leathers, "Garden Clubs Sponsor Ecology Symposium," *The Mission Hills Squire*, February 22, 1974, clipping in symposium notebook, WGCA.

21. Mildred S. Hoover, report, "How It All Began"; Lillian Diveley, report, "History of the Westport Garden Club"; Kessinger, "Festival Report," all in *The Westport Garden Club's Twenty-fifth Anniversary Program*, History Notebook, WGCA.

22. Vivian Foster, report, "Civic Projects," *The Westport Garden Club's Twenty-fifth Anniversary Program*, History Notebook, WGCA.

23. Josephine Stubbs, report, "Know Your G.C.A.," *The Westport Garden Club's Twenty-fifth Anniversary Program*, History Notebook, WGCA.

24. Diveley, "History of the Westport Garden Club."

Chapter 5
Educational Opportunities Expand

1. "GCA Statement of Purpose," Garden Club of America, https://www.gcamerica.org/about, accessed November 7, 2024.

2. Norma Sutherland, *President's Annual Report, June 1979*, WGCA.

3. Glenda-Jo Self, "Gardening This Week," *Kansas City Star*, June 5, 1981, clipping in Westport Garden Club Scrapbook: 1981–1985, WGCA.

4. "WGC's Strategic Plan Created for GCA," Long-Range Planning Committee Minutes May 6, 2013, minutes 2011–2013 folder, WGCA.

5. "WGC's Strategic Plan."

6. WGC Membership Meeting Minutes, April 2, 2018, minutes 2017–2018 folder, WGCA.

7. "The BoysGrow Program" and home page, BoysGrow, https://www.boysgrow.com/the-program/, accessed August 13, 2023.

8. "Conservation Report," March 11, 2004, minutes 2003–2004 folder, WGCA.

9. WGC Membership Meeting Minutes, October 4, 1976, Diana James, president's file, WGCA.

10. "Taberville Prairie Conservation Area," Missouri Department of Conservation, https://www.mdc.mo.gov/discover-nature/places/taberville-prairie-conservation-area, accessed June 26, 2023.

11. "Flint Hills," Flint Hills Discovery Center, https://www.flinthillsdiscovery.org, accessed July 23, 2023.

12. Helen Thompson, "Program Chair Report, 1985," Sally Wood, president's file, 1983–1985, WGCA.

13. Thompson, "Program Chair Report, 1985."

14. Marie Bell O'Hara, *President's Annual Report, June 1987*, WGCA.

15. "The Land Institute: Natural Systems Agriculture," The Land Institute, https://www.landinstitute.org/about-us/vision-mission, accessed July 20, 2023.

16. "Grand River Grasslands," Missouri Department of Conservation, https://www.mdc.mo.gov/your-property/priority-geographies/grand-river-grasslands, accessed July 20, 2023.

17. WGC Membership Meeting Minutes, April 6, 2015, minutes 2014–2015 folder, Nancy Lee Kemper, president's file, WGCA.

18. Kathy Gates, "One of the Greatest Migrations on Earth," *Bulletin*, fall 2014, Westport Garden Club Scrapbook: 2014–2015, WGCA.

19. "GCA's First National Medal in Photography," The Garden Club of America, https://www.gcamerica.org/news/get?id=3037, accessed November 6, 2024; "Arbor Day Farm," Arbor Day Farm, https://www.arbordayfarm.org/about/, accessed October 12, 2023.

20. "Advice Mixes with Charm," *Kansas City Star*, April 16, 1976, clipping in Westport Garden Club Scrapbook: 1975–1976, WGCA.

21. Norma Sutherland, *President's Annual Report, June 1979*, WGCA.

22. "Awards: Remembering 1982 Honorary Member, Dr. Ruth Patrick," The Garden Club of America, November 30, 2018, https://www.gcamerica.org/members/news/get?id=1366.

23. Laura Rollins Hockaday, "Designs of Delight," *Kansas City Star*, May 1, 1986, clipping in Sally Cowherd Memorial Lecture Series folder 1986, WGCA.

24. Dorothy Halsey Hughes, "Sally Cowherd Lecture—1986"; invitation, both in Sally Cowherd Memorial Lecture Series folder 1986, WGCA.

25. Dee Hughes and Hazel Barton, "Sally Cowherd Memorial Lecture Series," Sally Cowherd Memorial Lecture Series folder 1988, WGCA.

26. Mrs. Thomas B. [Margaret] Hall III and Mrs. Lathrop M. [Dody] Gates, "Project Planning Report, 1992–1993, Westport Garden Club," Marilyn Patterson, president's file, WGCA.

27. Sallie Bet Watson, "Sally Cowherd Memorial Lecture," Sally Cowherd Memorial Lecture Series folder 1993, WGCA.

Chapter 6
Flower Shows Promote the Love of Gardening

1. Martha Wilcox Van Allen and Bliss Caulkins Clark, eds., *Flower Show and Judging Guide*, preface.

2. *Westport Garden Club Annual Report, 1972–1973*, WGC Annual Reports, 1960s–1970s, WGCA.

3. Norma Sutherland, *President's Annual Report, June 1979*, WGCA.

4. "The Westport Garden Club Flower Show Schedule," July 8, 1980, WGCA.

5. Richard R. Callahan to Mrs. William P. [Florence] Kline, April 22, 1985, "Westport Garden Club Flower Festival 1985: Chairman's Report," WGCA.

6. Marc F. Wilson to Barbara Seidlitz, August 16, 1985, "Flower Festival 1985: Chairman's Report," WGCA.

7. Sallie Watson and Sally Wood, "Zone XI Meeting Report," April 1987, Marie Bell O'Hara, president's file, 1985–1987, WGCA.

8. Minutes of Special Evaluation Meeting of the Flower Festival, May 23, 1990, Suzanne Vawter, president's file, 1989–1991, WGCA.

9. "Report of Chairman and Co-Chairman of the 1990 Flower Festival," WGCA.

10. *Horticulture and Flower Show: "Past Is Prologue,"* booklet, WGCA.

11. Hazel Barton, *President's Annual Report, June 1995*, WGCA.

12. *"Joie de Vivre," A Garden Club of America Flower Show*, booklet, 27, WGCA.

13. Laura Powell and Peggy Rooney, "Flower Show Report: 'Fleur-ishing Art: An American Impression,'" April 23, 2015, WGCA.

14. *The Enduring Spirit and Traditions of The Kansas City Country Club, 1896–2021*, 72.

15. Candid photo of Norma Sutherland and Susie Vawter, with cutline, *The Independent*, May 1, 1993.

Chapter 7
Partnerships Pay Big Dividends

1. "GCA Statement of Purpose," The Garden Club of America, https://www.gcamerica.org/about, accessed November 7, 2024.

2. Hazel Barton, *President's Annual Report, June 1995*, WGCA.

3. Laura Kemper Carkener, *President's Annual Report, June 1967*, WGCA.

4. *The Westport Garden Club Conservation Committee Annual Report, 1995–1996*, Margaret Hall, president's file, 1995–1997, WGCA.

5. Barbara Seidlitz, *President's Annual Report, June 1981*, WGCA.

6. WGC Membership Meeting Minutes, September 9, 2013, minutes 2013–2014 folder, WGCA.

7. "Park Board," Mission Hills, Kansas, https://www.missionhillsks.gov/34/Park-Board, accessed May 29, 2023; Lorelei Gibson, interview by author, May 28, 2023.

8. Barbara Seidlitz, *Mission Hills Beautification Committee Annual Report, 1986–1987*, WGCA; Minutes, October 27, 2004, Paget Higgins, president's file, 2003–2205, WGCA.

9. Sallie Bet Watson, *President's Annual Report, June 1988*; Sally Wood, *President's Annual Report, June 1984*, both in WGCA.

10. Michael S. Churchman to Betty Thompson, July 13, 1995, Annual Reports, 1994–1995, WGCA.

11. Norma Sutherland, "President's Report, Zone XI Meeting, Chicago, June 27–29, 1977," president's file, 1977–1979; Penny Selle, letter to WGC members, June 24, 1987, Marie Bell O'Hara, president's file, 1985–1987, both in WGCA.

12. Boots Churchill, "Spang Memorial: A Family's Tribute"; "Spang Memorial," WGC Membership Meeting Minutes, October 3, 2022, minutes 2022–2024 folder, both in WGCA.

13. "The WGC Year-End Gifting," April 1, 2013, minutes 2011–2013 folder, WGCA.

14. "Loose Park," Kansas City Parks and Recreation, https://www.kcparks.org/places/loose-park/, accessed December 4, 2022.

15. "The Laura Conyers Smith Municipal Rose Garden," in Kansas City Parks and Recreation, "Loose Park," printout in Rose Garden folder, Sally West, president's file, 2005–2007, WGCA.

16. Virginia S. Foresman, "Board of Directors Meeting, May 1, 1981," Barbara Seidlitz, president's file, 1979–1981, WGCA.

17. Virginia McCanse to author, email, September 18, 2023.

18. Ginny McCanse, Jo Missildine, Molly Fusselman, Eileen McManus, and Dave Patton, *Explorers' Field Guide: Exploring and Observing Loose Park's Arboretum*.

19. Paget Higgins, *President's Annual Report, June 2004*, WGCA.

20. Executive Committee Meeting Minutes, November 14, 2005, Sally West, president's file, 2005–2007, WGCA.

21. "The History of Powell Gardens," Powell Gardens, https://www.powellgardens.org/history/, accessed April 11, 2023.

22. Wendy Powell to author and Eric Tschanz, email, April 10, 2023.

23. Powell to author and Eric Tschanz.

24. "Long Range Planning Community Donation Study," May 2020, Carolyn Kroh, president's file, 2019–2021, WGCA; *"Joie de Vivre"* booklet, 27.

25. "Long Range Planning Community Donation Study"; WGC Membership Meeting Minutes, May 5, 2013, minutes 2011–2013 folder, WGCA.

26. "Our Mission," Kansas City Community Gardens, https://www.kccg.org/history-mission, accessed June 25, 2023.

27. "Beanstalk Children's Garden," Kansas City Community Gardens, https://www.kccg.org/beanstalk-childrens-garden, accessed June 26, 2023.

28. Sally West, *President's Annual Report, June 2007*, WGCA.

29. WGC Membership Meeting Minutes, June 6, 2023, minutes 2022–2024 folder, WGCA.

30. "About Us: Our Story," The Giving Grove, https://www.givinggrove.org/about-us, accessed July 20, 2023.

31. Home page, Kansas City Community Gardens, https://www.kccg.org, accessed June 25, 2023.

32. "Philanthropy: Deep Roots," The Westport Garden Club, https://www.thewestportgardenclub.org/philanthropy, accessed October 6, 2022; WGC Native Plant Task Force Meeting Minutes, April 12, 2014, minutes 2014–2015 folder, WGCA.

33. WGC Membership Meeting Minutes, September 16, 2014, minutes 2014–2015 folder, WGCA.

34. WGC Membership Meeting Minutes, December 12, 2017, minutes 2017–2018 folder, WGCA; "2023 National GCA Medalists Announced," "Roy Diblik," The Garden Club of America, January 9, 2023, https://www.gcamerica.org/news/get?id=4099.

35. Minutes, December 12, 2017.

36. "Deep Roots KC," Deep Roots, https://www.deeproots.org/about, accessed October 3, 2022.

37. WGC Annual Meeting Minutes, June 7, 2017, minutes 2016–2017 folder, WGCA.

38. "2023 National GCA Medalists Announced," "Robert J. Berkebile," The Garden Club of America, January 9, 2023, https://www.gcamerica.org/news/get?id=4099.

39. Ginny McCanse, "Partners for Plants Project: History and Report," September 9, 2019, Carolyn Kroh, president's file, 2019–2021.

40. McCanse, "Partners for Plants Project."

41. McCanse, "Partners for Plants Project"; Ginger Werp to author, emails, December 10, 2022; August 14, 2023.

42. "P4P: Jerry Smith Native Prairie Park and Kansas City WildLands Partnership," The Garden Club of America, https://www.gcamerica.org/members:project/details/id/33, accessed April 17, 2022.

Chapter 8
Supporting Our Partners and Funding Our Projects

1. The Westport Garden Club Directory, 2022–2023, 2.

2. Lillian Diveley, report, "History of The Westport Garden Club," 14, The Westport Garden Club's Twenty-fifth Anniversary Program, History Notebook, WGCA; Mrs. Frederic [Diana] James, "A Garden Room Within the Nelson Gallery," in Diana Morgan Olcott, Winds of Change, 1963–1988: 75th Anniversary Chronicle, 141–42.

3. Laurence Sickman, "Comments of the Director," Annual Report 1976, NAMAA.

4. Sickman, "Comments of the Director"; Tara Laver to author, email, October 28, 2021; Elizabeth Hull, Last Will and Testament, December 21, 1994, Ridgefield Garden Club Archives, courtesy of Terry McManus.

5. Norma Sutherland, "President's Report, Zone XI Meeting, Chicago, June 27–29, 1977," president's file, 1977–1979; Barbara Seidlitz, President's Annual Report, June 1980, both in WGCA.

6. Sutherland, "President's Report, Zone XI Meeting"; WGC Membership Meeting Minutes, December 9, 2003, Paget Higgins, president's file, 2003–2005, WGCA.

7. Mrs. Benjamin C. [Libby] Adams, "An Irish Cottage," Irish cottage notebook, WGCA.

8. "The National Crittenton Mission," Crittenton, https://www.crittentonsocal.org/crittenton-history/, accessed October 22, 2022.

9. Norma Sutherland, President's Annual Report, June 1979, WGCA.

10. "To the Family of Evert Asjes, Jr.," April 27, 1989, flower festivals folder, WGCA; "About Us," Rosehill Gardens, https://www.rosehillgardens.com/about-us/, accessed December 22, 2022.

11. Mrs. G. Guyton [Laura] Carkener, "Gardening Is Teen-Age Therapy," Bulletin, April 1980; Sutherland, President's Annual Report.

12. Glenda-Jo Self, "Gardening This Week," Kansas City Star, April 18, 1980, clipping in Westport Garden Club Scrapbook: 1979–1981, WGCA; Virginia Foresman, WGC Membership Meeting Minutes, February 2, 1981, Barbara Seidlitz, president's file, 1979–1981, WGCA.

13. WGC Membership Meeting Minutes, February 2, 1981, and May 4, 1981, Barbara Seidlitz, president's file, 1979–1981; Barbara Seidlitz, President's Annual Report, June 1981; all in WGCA.

14. "The Westport Garden Club," The Independent, April 9, 1983; Sally Wood, President's Annual Report, June 1984, WGCA.

15. Horticulture Chairman's Annual Report, 2008–2009, Ann Readey, president's file, 2007–2009, WGCA.

16. Krista Allen to Helen Lea, email, August 20, 2010, Blair Hyde, president's file, 2009–2011, WGCA.

17. Betty Goodwin, "The WGC Project 2000," November 1998, Children's Center folder, WGCA.

18. WGC Membership Meeting Minutes, October 6, 2003, Paget Higgins, president's file, 2003–2005, WGCA; Wendy Powell to Cale, Victoria, and Mary Lynne, November 30, 2003, Children's Center folder, WGCA.

19. Wendy Powell to the Children's Center Project Committee, January 26, 2004, Children's Center folder, WGCA; Wendy Powell, "Children's Center Report," WGC Membership Meeting Minutes, February 2, 2004, Paget Higgins, president's file, WGCA.

20. "The Tulip Project Surpasses Its 10,000 Tulip Goal," Tulips on Troost folder, Sally West, president's file, 2005–2007, WGCA.

21. Dennis Boone, "After the Storm, Joplin Means Business," Ingram's, June 2011, clipping in Joplin Project folder, WGCA.

22. Jill Bunting to Betty Kessinger, email, August 17, 2011; Betty Kessinger to Wendy Powell, email, June 2, 2011, both in Joplin Project file, WGCA.

23. "Project Joplin," WGC Minutes, September 12, 2011, minutes 2011–2013 folder, WGCA; Betty Kessinger, President's Annual Report, June 2012, WGCA.

24. Diana Kline to Betty Kessinger, email, February 27, 2013, in Joplin Project folder; WGC Annual Meeting Minutes, June 5, 2013, minutes 2013–2015 folder, both in WGCA.

25. "Mission and Vision," Kansas City Museum, https://www.kansascitymuseum.org/about/, accessed July 1, 2023.

26. Sharon Orr to author, email, June 11, 2023.

27. Betty Kessinger, Annual Meeting Minutes, June 6, 2012, minutes 2011–2013 folder; Christopher Leitch to Betty Kessinger, email, June 26, 2012, Betty Kessinger, president's file, 2011–2013, both in WGCA.

28. Betty Kessinger, "A Short Report on the State of The Westport Garden Club," WGC Annual Meeting Minutes, June 6, 2012, minutes 2011–2013 folder, WGCA.

29. "Philanthropy: Deep Roots," The Westport Garden Club, https://www.thewes tportgardenclub.org/philanthropy, accessed October 6, 2022.

30. "2015 Monarch Butterfly Conservation Fund Grant Slate," National Fish and Wildlife Foundation, https://www.nfwf.org>monarchgrants15-0925, accessed October 4, 2022.

31. "Loose Park Native Plant Demonstration Gardens," Grow Native!/Missouri Prairie Foundation, https://www.grownative.org/project/loose-park-native-plant-demonstration -gardens/, accessed October 20, 2022.

32. "Loose Park Native Plant Demonstration Gardens."

33. Wendy Powell to author, email, July 15, 2023.

34. WGC Membership Meeting Minutes, April 3, 2023, minutes 2022–2024 folder, WGCA.

35. WGC Minutes, January 5, 2004, Paget Higgins, president's file, WGCA; Ann Readey, *President's Annual Report, June 2008*, WGCA.

36. "Entertaining Gardens," letter to prospective patrons, Betty Kessinger, president's file, 2011–2013; Membership Meeting Minutes, May 6, 2013, minutes 2011–2013 folder, WGCA.

37. Membership Meeting Minutes, November 6, 2017, minutes 2017–2018 folder, WGCA.

Chapter 9
It's All About Gardens

1. "GCA Statement of Purpose," The Garden Club of America, https://www.gcameri ca.org/about, accessed November 7, 2024.

2. Norma Sutherland, "Flower of the Week."

3. WGC Membership Meeting Minutes, January 8, 2018, minutes 2017–2018 folder, WGCA.

4. Carolyn Kroh to author, email, August 30, 2023.

5. Margaret Hall to author, email, August 30, 2023.

6. WGC Membership Meeting Minutes, September 11, 2017, minutes 2017–2018 folder, WGCA; "The Healthy Yard Pledge," The Garden Club of America, https://www .gcamerica.org/members/committees-conservation#The-Healthy-Yard-Pledge, accessed October 5, 2023.

7. "Governor Parson Proclaims April 2023 to Be Native Plant Month," Missouri Governor Michael L. Parson, https://www.governor.mo.gov/proclamations/governor-parson -proclaims-april-2023-to-be-native-plant-month, accessed November 15, 2024; Marilyn Donahue, (GCA) President's Letter, *Bulletin*, Fall 2023, 3.

8. "GCA Medals: Cynthia Pratt Laughlin Medal," The Garden Club of America, https://www.gcamerica.org/awards/details/id/13, accessed November 15, 2024; "Bob Berkebile," The Garden Club of America, https://www.gcamerica.org/members:videos /details?id=582, accessed October 5, 2023.

9. "Visiting Gardens," The Garden Club of America, https://www.gcamerica.org /index-cfm/members/visiting-gardens-manual, accessed July 13, 2023.

10. "History of the Garden History and Design Committee," The Garden Club of America, https://www.gcamerica.org/membersonly/docs/GH&D, accessed October 27, 2024; "Archives of American Gardens," Smithsonian Institution, https://www.si.edu /siasc/american_gardens, accessed May 27, 2023.

11. "Autochromes: Frank Lauder Collection," Kansas City Public Library, https: //www.kchistory.org/collection/autochromes-frank-lauder-collection; "A Step Back in Time," *Star Magazine*, November 5, 2000; Boots Leiter to Anita Robb, August 21, 2000, envelope in GH&D notebook, WGCA.

12. Joyce Connolly, Joyce's 'Worthy Gardens' letter, November 9, 2009, GH&D notebook, WGCA.

13. Barbara Kohoe to Boots Leiter, email, January 1, 2006; Jan Riss, "Gracie's Garden," Gracie's Garden folder, GH&D notebook, WGCA.

14. Lorelei Gibson, interview by author, May 28, 2023.

15. Lorelei Gibson, interview by author, May 28, 2023.

16. Cort Sinnes, "A Salute to Sarah and Virginia Weatherly," Sally Cowherd Memorial Lecture, November 2, 1989, in Sally Cowherd Memorial Lecture Series folder; George Gurley, "The Garden of Virginia and Sarah Weatherly," editorial, *Kansas City Star*, n.d., clipping in Weatherly Garden folder, GH&D notebook, WGCA.

17. Nancy Lee Kemper, *Garden History and Design Committee Annual Report, 2009–2010*, Blair Hyde, president's file, 2009–2011; Joyce Connolly to Ann Hosford, September 25, 2014, Reed Dillon folder, GH&D notebook, both in WGCA.

18. Kelly Crawford to Mr. and Mrs. Gibson, November 30, 2015, Lorelei Gibson Private Collection.

19. "Marvin K. Snyder" (obituary), Legacy, https://www.legacy.com/us/obituaries /name/marvin-snyder-obituary?id=5697054, accessed April 15, 2023.

20. "November 2021 Featured Garden," The Garden Club of America, https://www .gcamerica.org/members:news/printnews/id/3585, accessed April 15, 2023.

Chapter 10
"Everything IS Up-to-Date in Kansas City!"

1. Marilyn Patterson, *President's Annual Report, June 1993*; Executive Board Meeting Minutes, July 9, 2011, minutes 2011–2013 folder, both in WGCA.

2. Margaret Hall to author, email, November 5, 2023.

3. WGC Membership Meeting Minutes, May 3, 2004, Paget Higgins, president's file, 2003–2005, WGCA.

4. "Annual Meeting 2005—Award Winners," *Bulletin*, August–September 2005, 15; "Cherie Sutton Pettit" (obituary), *San Francisco Chronicle*, March 25–26, 2008.

5. Helen Lea to author, email, May 12, 2024.

6. WGC Membership Meeting Minutes, November 1, 2004, Paget Higgins, president's file, 2003–2005, WGCA; Lorelei Gibson, "Nomination for a Certificate of Acknowledgment: City of Mission Hills," and The Garden Club of America, "GCA Club Civic Improvement Award," June 2007, both in Lorelei Gibson Private Collection.

7. Norma Sutherland, "Those Plant Exchange Babies."

8. Sallie Bet Watson, *President's Annual Report, June 1988*, WGCA; DeeDee Adams, presentation at WGC meeting, February 6, 2023.

9. Katherine G. Gates and Ellison B. Lambert, "Westport Garden Club Annual Meeting, Horticulture Committee," June 7, 1993, Marilyn Patterson, president's file, 1991–1993, WGCA.

10. WGC Membership Meeting Minutes, May 3, 2004, Paget Higgins, president's file, 2003–2005, WGCA.

11. WGC Annual Meeting Minutes, June 8, 2005, Paget Higgins, president's file, 2003–2005, WGCA.

12. "Annual Meeting 2005—Award Winners," *Bulletin*, August–September 2005, 15.

13. WGC Annual Meeting Minutes, June 8, 2005, Paget Higgins, president's file, 2003–2005, WGCA.

14. "Everything IS Up-to-Date in Kansas City!" *Bulletin*, August–September 2005.

15. WGC Membership Meeting Minutes, September 12, 2005, Sally West, president's file, 2005–2007, WGCA.

Chapter 11
Awards: Treasured Recognition for Outstanding Work

1. "GCA Awards," The Garden Club of America, https://www.gcamerica.org/gca-awards, accessed November 17, 2024.

2. *The Westport Garden Club Directory, 2022–2023*, 38.

3. *The Westport Garden Club Directory, 2022–2023*, 38; Frederick Kirschenmann, PhD, to Mrs. Robert D. Gongaware, January 21, 2011, in Wes Jackson 2012 folder, WGCA; "About the Land Institute," The Land Institute, https://www.landinstitute.org/about-us/, accessed July 3, 2023.

4. "Conservation Heroes," *ConWatch*, Summer 2023.

5. "Elizabeth Abernathy Hull Award," The Garden Club of America, https://www.gcamerica.org/members/civic-improvement-manual, accessed July 30, 2023.

6. WGC Membership Meeting Minutes, October 2, 2023, minutes 2022–2023 folder, WGCA.

7. "The Creative Leadership Award," The Garden Club of America, accessible to members at https://www.gcamerica.org/members/zone/cla-instructions, accessed November 17, 2024.

8. "Zone Horticulture Commendation, 2013: Alan Branhagen," The Garden Club of America, accessible to members at https://www.gcamerica.org/members:awards/winners search, accessed November 17, 2024.

9. Ann P. Readey to Mrs. Frederick M. Goodwin Jr., Chairman, GCA Garden History and Design Committee, November 26, 2007, "Nutterville" folder, WGCA.

10. "James B. Nutter Sr." (obituary), Dignity Memorial, https://www.dignitymemorial.com/obituaries/kansas-city-mo/james-nutter-7476733, accessed July 3, 2023.

11. WGC Annual Meeting Minutes, June 6, 2023, minutes 2020–2022 folder, WGCA.

Bibliography

Works Cited

Arbor Day Farm. "Arbor Day Farm." https://www.arbordayfarm.org/about/. Accessed October 12, 2023.

Battles, Marjorie Gibbon, and Catherine Colt Dickey. *Fifty Blooming Years, 1913–1963*. The Garden Club of America, 1963.

Biography.com Editors and Tyler Piccotti. "Lady Bird Johnson." Updated November 13, 2023. https://www.biography.com/history-culture/lady-bird-johnson.

Boone, Dennis. "After the Storm, Joplin Means Business." *Ingram's*, June 2011.

BoysGrow. "The BoysGrow Program." https://www.boysgrow.com/the-program/. Accessed August 13, 2023.

Bridging the Gap. "KC Metro Champion Trees." https://www.bridgingthegap.org/kc-metro-champion-trees.

Bulletin. "Annual Meeting." No. XXV, May 1918.

Bulletin. "Annual Meeting 2005—Award Winners." August–September 2005.

Bulletin. "Everything IS Up-to-Date in Kansas City!" August–September 2005.

Bulletin. "National Gardens." No. III, January 1914.

Bulletin. "National Gardens." No. IV, April 1914.

Carkener, Mrs. G. Guyton [Laura]. "Gardening Is Teen-Age Therapy." *Bulletin*, April 1980.

Carson, Rachel. *Silent Spring*. Greenwich, CT: Fawcett, 1962.

Caudill, Claire P., Betty Pinkerton, Diane B. Stoner, and Dede Petri. *A History of Conservation and National Affairs and Legislation: The Garden Club of America, 1913–2019*. Rev. ed. The Garden Club of America, 2019. Accessible to members at https://www.gcamerica.org/members:publications/details/id/28.

Churchill, Boots. "Spang Memorial: A Family's Tribute." *La Belle Jardinière*, Fall 2022.

Crittenton Services for Children and Families. "The National Crittenton Mission." https://www.crittentonsocal.org/crittenton-history. Accessed October 22, 2022.

Cullen, Betsy. "Elizabeth Abernathy Hull." In *Remember the Ladies: Notable Women of Ridgefield*, 29–35. Ridgefield Historical Society, 2008.

Deep Roots. "Deep Roots KC." https://www.deeproots.org/about. Accessed October 3, 2022.

Dignity Memorial. "James B. Nutter Sr." (obituary). https://www.dignitymemorial.com/obituaries/kansas-city-mo/james-nutter-7476733. Accessed July 3, 2023.

Dodd, Monroe. "How Floods Shaped the Kansas City We Know Today." KCUR 89.3, NPR in Kansas City, August 10, 2015. https://www.kcur.org/arts-life/2015-08-10/how-floods-shaped-the-kansas-city-we-know-today. Accessed May 5, 2020.

The Enduring Spirit and Traditions of The Kansas City Country Club, 1896–2021. Brookfield, MO: Donning Company, 2021.

Flint Hills Discovery Center. "Flint Hills." https://www.flinthillsdiscovery.org. Accessed July 23, 2023.

The Garden Club of America. "Awards: Remembering 1982 Honorary Member, Dr. Ruth Patrick." November 30, 2018. Accessible to members at https://www.gcamerica.org/members/news/get?id=1366.

The Garden Club of America. "Bob Berkebile." Accessible to members at https://www.gcamerica.org/members:videos/details?id=582, accessed October 5, 2023.

The Garden Club of America. "Conservation Heroes." *ConWatch*, Summer 2023.

The Garden Club of America. "Elizabeth Abernathy Hull Award." Accessible to members at https://www.gcamerica.org/members/civic-improvement-manual. Accessed July 30, 2023.

The Garden Club of America. "GCA's First National Medal in Photography." https://www.gcamerica.org/news/get?id=3037. Accessed October 12, 2023.

The Garden Club of America. "The Healthy Yard Pledge." Accessible to members at https://www.gcamerica.org/members/committees-conservation#The-Healthy-Yard-Pledge. Accessed October 5, 2023.

The Garden Club of America. "November 2021 Featured Garden." Accessible to members at https://www.gcamerica.org/members:news/printnews/id/3585. Accessed April 15, 2023.

The Garden Club of America. "P4P: Jerry Smith Native Prairie Park and Kansas City WildLands Partnership." Accessible to members at https://www.gcamerica.org/members:project/details/id/33. Accessed April 17, 2022.

The Garden Club of America. "2023 National GCA Medalists Announced," "Robert J. Berkebile." January 9, 2023. https://www.gcamerica.org/news/get?id=4099.

The Garden Club of America. "2023 National GCA Medalists Announced," "Roy Diblik." January 9, 2023. https://www.gcamerica.org/news/get?id=4099.

The Garden Club of America. "Visiting Gardens." https://www.gcamerica.org/index-cfm/members/visiting-gardens-manual. Accessed July 13, 2023.

The Garden Club of America. "Zone Horticulture Commendation, 2013: Alan Branhagen." Accessible to members at https://www.gcamerica.org/members:awards/winnerssearch. Accessed November 17, 2024.

The Garden Club of Philadelphia. "About Us." https://www.thegardenclubofphiladelphia.org/about. Accessed June 6, 2020.

Gates, Kathy. "One of the Greatest Migrations on Earth." *Bulletin*, Fall 2017.

The Giving Grove. "About Us: Our Story." https://www.givinggrove.org. Accessed July 20, 2023.

Goodman, Ernestine Abercrombie. *The Garden Club of America: History 1913–1938*. Philadelphia: Edward Stern and Co., 1938.

Grow Native!/Missouri Prairie Foundation. "Loose Park Native Plant Demonstration Gardens." https://www.grownative.org/project/loose-park-native-plant-demonstration-gardens/. Accessed October 20, 2022.

Hockaday, Laura Rollins. "Designs of Delight." *Kansas City Star*, May 1, 1986.

The Independent. "Beauty Keynotes Luncheon." December 17, 1960.

The Independent. "Busy Days for Club." September 21, 1957.

The Independent. Candid photo of Norma Sutherland and Susie Vawter, with cutline. May 1, 1993.

The Independent. Candid photo of Mrs. Mason L. Thompson, with cutline. July 1, 1950.

The Independent. "Club Season Ends Brilliantly." June 17, 1961.

The Independent. "Flower Festival." May 26, 1956.

The Independent. "Flower Festival." February 4, 1961.

The Independent. "Flower Festival at Gallery." May 28, 1955.

The Independent. "Garden Tour by Westport Garden Club." June 16, 1962.

The Independent. "The Ivy Eagle." June 16, 1962.

The Independent. "The Lakeside Nature Center." May 2, 1970.

The Independent. "Legends of Christmas." December 12, 1987.

The Independent. "Nelson Gallery to Host 4,000." June 13, 1960.

The Independent. "Preparations for Sixth Flower Festival." April 20, 1963.

The Independent. "Project No. 1." April 4, 1952.

The Independent. "Rozzelle Court of Nelson Gallery." July 2, 1966.

The Independent. "Third Annual Flower Festival." April 6, 1957.

The Independent. "A Tradition Continues at the Nelson Gallery." April 5, 1969.

The Independent. "The Westport Garden Club." April 9, 1983.

The Independent. "Westport Garden Club at Sleepy Hill Farm." July 21, 1951.

The Independent. "Yule Customs on Display." December 16, 1961.

The Independent. "Zone Meeting." October 12, 1957.

Introduction to the Linda Hall Library Arboretum (brochure). Linda Hall Library, n.d.

The Johnson County Sun. "Westport Garden Club Gives $200 to Northwest." May 31, 1974.

Kansas City Community Gardens. "Beanstalk Children's Garden." https://www.kccg.org/beanstalk-childrens-garden. Accessed June 26, 2023.

Kansas City Community Gardens. Home page. https://www.kccg.org. Accessed June 25, 2023.

Kansas City Community Gardens. "Our Mission." https://www.kccg.org/history-mission. Accessed June 25, 2023.

Kansas City Museum. "Mission and Vision." https://www.kansascitymuseum.org/about/. Accessed July 1, 2023.

Kansas City Parks and Recreation. "Loose Park." https://www.kcparks.org/places/loose-park. Accessed December 4, 2022.

Kansas City Public Library. "Autochromes: Frank Lauder Collection." https://www.kchistory.org/collection/autochromes-frank-lauder-collection. Accessed May 9, 2024.

Kansas City Star. "About 300 at Opening at New Junior Gallery." January 30, 1960.

Kansas City Star. "Advice Mixes with Charm." April 16, 1976.

Kansas City Star. "Busily at Work on Their Major Landscaping Effort of the Season." April 1, 1952.

Kansas City Star. "Control on Tree Men." December 31, 1951.

Kansas City Star. "Fashions for Garden Parties and Garden Digging." April 23, 1959.

Kansas City Star. "Primitive Art Show Opening Is Social Event." January 21, 1962.

Kansas City Times. "Medal to Garden Club for Junior Art Gallery." May 21, 1964.

Kansas Historical Society. "Flood of 1951." *Kansapedia*. June 2011. https://www.kshs.org/kansapedia/flood-of-1951/17163.

The Land Institute. "About the Land Institute." https://www.landinstitute.org/about-us/. Accessed July 20, 2023.

Lawrence Weekly World. "Col. Abernathy Dead." December 18, 1902.

Leathers, Tom. "Garden Clubs Sponsor Ecology Symposium." *The Mission Hills Squire*, February 22, 1974.

Legacy. "Marvin K. Snyder" (obituary). https://www.legacy.com/us/obituaries/kansascity/name/marvin-snyder-obituary. Accessed April 15, 2023.

Linda Hall Library. "About Linda Hall Library." https://www.lindahall.org/about/our-story. Accessed May 2, 2020.

Linda Hall Library. "Introduction to the Linda Hall Library Arboretum." https://www.lindahall.org/arboretum.

Mack, Linda. *Madeline Island Summer Houses: An Intimate Journey*. I Was There Press, 2013.

McCanse, Ginny. *Explorers' Field Guide: Exploring and Observing Loose Park's Arboretum*. With contributions by Molly Fusselman, Eileen McManus, and Dave Patton. Garden Center Association of Greater Kansas City and Board of Parks and Recreation Commissioners, Kansas City, Missouri, 2008.

Mission Hills, Kansas. "Park Board." https://www.missionhillsks.gov/34/Park-Board. Accessed May 29, 2023.

Missouri Department of Conservation. "Grand River Grasslands." https://www.mdc.mo.gov/your-property/priority-geographies/grand-river-grasslands. Accessed July 20, 2023.

Missouri Department of Conservation. "Taberville Prairie Conservation Area." https://www.mdc.mo.gov/discover-nature/places/taberville-prairie-conservation-area. Accessed June 26, 2023.

Missouri Governor Michael L. Parson. "Governor Parson Proclaims April 2023 to Be Native Plant Month." https://www.governor.mo.gov

/proclamations/governor-parson-proclaims-april-2023-to-be
-native-plant-month. Accessed November 15, 2024.

National Fish and Wildlife Foundation. "2015 Monarch Butterfly
Conservation Fund Grant Slate." https://www.nfwf.org/sites/de
fault/files/monarch/Documents/monarchgrants15-0925.pdf.
Accessed October 4, 2022.

National Women's History Museum. "Gardening Clubs: Fertile Ground
for Women's Activism." April 16, 2017. https://www.womenshis
tory.org/articles/gardening-clubs.

Olcott, Diana Morgan, comp. *Winds of Change, 1963–1988: 75th
Anniversary Chronicle*. Princeton: The Garden Club of America,
1988.

Powell Gardens. "The History of Powell Gardens." https://www.powellgar
dens.org/history/. Accessed April 11, 2023.

Rosehill Gardens. "About Us." https://www.rosehillgardens.com/about
-us/. Accessed December 22, 2022.

San Francisco Chronicle. "Cherie Sutton Pettit" (obituary). March 25–26,
2008.

Scenic America. "Scenic America Remembers Marion Fuller Brown."
September 18, 2019. https://www.scenic.org/2019/09/18/sce
nic-america-remembers-marion-fuller-brown.

Seale, William. *The Garden Club of America: 100 Years of a Growing
Legacy*. Washington, DC: Smithsonian Books, 2012.

Self, Glenda-Jo. "Gardening This Week." *Kansas City Star*, April 18, 1980.

Self, Glenda-Jo. "Gardening This Week." *Kansas City Star*, June 5, 1981.

Seidelman, James. "A Junior Gallery Within a Large Museum." *Museum
News*, April 1961.

Smithsonian Institution. "Archives of American Gardens." https://www
.si.edu/siasc/american_gardens. Accessed May 27, 2023.

Star Magazine. "A Step Back in Time." November 5, 2000.

Sutherland, Norma. "Flower of the Week." *Bulletin*, May 1985.

Sutherland, Norma. "Those Plant Exchange Babies." *Bulletin*, October
1981.

Van Allen, Martha Wilcox, and Bliss Caulkins Clark, eds. *Flower Show
and Judging Guide*. Rev. ed. The Garden Club of America, 2009.

The Westport Garden Club. "Philanthropy: Deep Roots." https://www
.thewestportgardenclub.org/philanthropy. Accessed October 6,
2022.

The Westport Garden Club. *The Westport Garden Club Directory, 2022–
2023*. The Westport Garden Club, 2023.

The Westport Garden Club. *The Westport Garden Club Directory, 2023–
2024*. The Westport Garden Club, 2024.

Manuscript/Record Collections

The Garden Club of America Archives (GCA Archives). New York, NY.

The Garden Club of Philadelphia Archives (GCPA). The Historical
Society of Pennsylvania. Philadelphia, PA.

The Independent (magazine) Archives. Prairie Village, KS.

The Nelson-Atkins Museum of Art Archives (NAMAA). Kansas City, MO.

Annual Reports

Department of Education Records, RG 32

Ephemera Collection, RG 70

Nelson-Atkins Media Services

Westport Garden Club Records, RG 71/01

William Rockhill Nelson Trust, Trustees' Minutes

Ridgefield Garden Club Archives. Ridgefield, CT. Access courtesy of Terry
McManus.

Hull, Elizabeth, Last Will and Testament, December 21, 1994

"Memories of Miss Hull," May 2023

Miscellaneous articles and documents

The Westport Garden Club Archives (WGCA). Linda Hall Library,
Kansas City, MO.

Annual Reports

Articles of incorporation folder

Children's Center folder

Flower festivals folder

Flower show booklets and reports

Garden History and Design Committee notebook

Gracie's Garden folder

History Notebook

Irish cottage notebook
Joplin Project folder
La Belle Jardinière magazines
Minutes of membership meetings, board meetings, Executive Committee
 meetings
"Nutterville" folder
Presidents' notebooks and files
Reed Dillon folder
Rose Garden folder
Sally Cowherd Memorial Lecture Series folder
Scrapbooks
Symposium notebook
Tulips on Troost folder
Weatherly Garden folder
Wes Jackson 2012 folder
Zone meeting booklets
Wisconsin Historical Society Archives (WHSA). Madison, WI.
Elizabeth Abernathy Hull Collection

Private Collections
Gibson, Lorelei. Private Collection.
Hall, Margaret. Private Collection.

Interviews
With many members of The Westport Garden Club, including the following:
Adams, DeSaix. Interviews by author. January 3, 2022; February 6, 2023,
 Kansas City, MO.
Gibson, Lorelei. Interview by author. May 28, 2023, Mission Hills, KS.
Grace, Laura Lee Carkener. Interview by author. June 8, 2023, Mission
 Hills, KS.
Hyde, Blair Peppard. Interview by author. June 8, 2022, Kansas City,
 MO.

Emails
Barton, Bruce. Email to author, January 27, 2022.
Gates, Debbie Thompson. Emails to author, January 7, 2020; January 28,
 2020.
Hall, Margaret. Emails to author, August 30, 2023; November 5, 2023.
Kroh, Carolyn. Email to author, August 30, 2023.
Laver, Tara. Email to author, October 28, 2021.
Lea, Helen. Email to author, May 12, 2024.
McCanse, Virginia. Email to author, September 18, 2023.
McManus, Terry. Email to author, October 13, 2021; February 3, 2022;
 September 8, 2023.
Orr, Sharon. Email to author, June 11, 2023.
Powell, Wendy. Email to author, July 15, 2023.
Powell, Wendy. Email to Eric Tschanz and author, April 10, 2023.
Werp, Ginger. Emails to author, December 10, 2022; August 14, 2023.
Wilson, Marc. Email to author, April 10, 2024.